CHEMISTRY RESEARCH AND APPLICATIONS

NEW DEVELOPMENTS IN ANALYTICAL CHEMISTRY RESEARCH

CHEMISTRY RESEARCH AND APPLICATIONS

Additional books in this series can be found on Nova's website under the Series tab.

Additional e-books in this series can be found on Nova's website under the e-book tab.

CHEMISTRY RESEARCH AND APPLICATIONS

NEW DEVELOPMENTS IN ANALYTICAL CHEMISTRY RESEARCH

BREANNA GRANGER
EDITOR

nova publishers
New York

Additional color graphics may be available in the e-book version of this book.

Library of Congress Cataloging-in-Publication Data

ISBN: 978-1-63463-427-4

Published by Nova Science Publishers, Inc. † New York

CONTENTS

PREFACE

Carbon-based nanomaterials such as carbon nanotubes (CNT), carbon black and fullerenes are interesting materials for the development of analytical methods. In electrochemistry, carbon nanotubes improves the electron transfer and provides an improvement in the analytical sensitivity and selectivity of phenolic compounds determination. This book discusses several topics on analytical chemistry including enhanced electrochemical properties of multi-walled carbon nanotube paste electrode for simultaneous determination of phenolic compounds in the presence of surfactant; analytical methods for organic and inorganic toxics; bioinspired chemically modified electrodes for electroanalysis; and the adsorption properties of a new ion-selective hierarchically imprinted poly(acrylamide-co-egdma) for the extraction/pre-concentration of Co(II) from an aqueous medium.

Chapter 1 - The widespread use has made preservatives become a special monitored group in commercial cosmetics, especially in the *leave-on* products due to the prolonged exposure.

Additionally, recently there has been an increase in the demand for preservative-free cosmetics related to undesired adverse effects on human health, in order to minimize the amount in personal care products or produce preservative-free-cosmetics and verify the contents to avoid health risk.

Hence, the simultaneous separation and trace-level determination in personal care products are important for both merchandise reliability and quality assurance.

However, is not possible to exclude that heavy metals can be accumulated in the body over time and repetitive metal-containing product application can lead to significant exposure levels.

In the case of some metallic powders such as nickel (Ni), lead (Pb), and cadmium (Cd), they are unintentionally added to the cosmetics as impurities during manufacturing of an ingredient or product, and represent a minor source of exposure compared to other pathways such as water, food or air.

For these reasons it is necessary a deep knowledge of real toxics (organic and inorganic) content into personal care products and cosmetics using reliable, high-throughput, sensitive and selective analytical assays.

In this chapter, recently developed method for the extraction, separation, and quantitative determination of organic (phthalates and parabens) and inorganic toxic (heavy metals) in cosmetic products are reported.

Chapter 2 - This chapter aims to present main aspects of biomimetic sensors, their obtainment and analytical applications, briefly revisiting fundamentals of voltammetry. These fields have experienced an impressive growth in recent years; basic and applied researches on bioinspired chemically modified electrodes are opening, this way, a wide range of new specific applications. An electroanalytical biosensing device is a self-contained integrated unit capable of providing specific semi-quantitative, or even quantitative, analytical information by using a biological recognition element suitably retained in direct contact with an electrochemical transduction element. Such devices can be easily classified according to biological specificity-conferring mechanism, to physical-chemical signal transduction or to reaction category they monitor. The first well-established biomimetic sensor is the enzyme-based biosensor, in which different enzymes can be coupled to electrochemical platforms by many different ways in order to improve selectivity or/and sensitivity, such enzymes can be used as purified material or even in natural extracts, keeping the main advantages and reducing costs in many cases. Other interesting biological recognition mechanism exploited for electroanalytical purposes is antigen-antibody reactions, which leads to an immunosensor, a special kind of sensor in which the analytical device contains a biological material (antibody or antigen) immobilized onto its surface. Both described systems (enzyme-substrate and antigen-antibody) must be carefully assessed in order to achieve a responsive layer in which biochemical components can be entrapped without loss of activity or stability; in this sense many different immobilization approaches have been described. These bioelectroanalytical approaches are, in fact, limited to some natural or recombinant molecules, to the availability and stability of recognition elements and their capability of entrapment onto electrode surfaces, besides these costs of purified material are also a drawback. Aiming to overcome these limitations but keeping on mind the exploitation of biomolecular recognition principles, many studies have report the use of molecularly imprinted polymers (MIPs) as electrode modifier. The synthesis of such polymers may be reached from functional cross-linkable monomers in presence of a target/template molecule, being categorized according to the kind of occurring binding interactions (covalent or non-covalent) followed by a template removal step, vital for the recognition mechanism (since it allows the existence of complementary cavities). MIP-based sensors are in general, by means of sensitivity comparison, still inferior to biosensors. On the other hand, the great stability, low costs, and possibility of modeling different polymers based on many different templates; including molecules without natural biochemical recognition are still advantages that encourage scientists. Most of these applications are dependent on the related mass and charge transport properties. Thus, it becomes essential to understand the mechanisms of those transports in these materials, what can be carried out by electrochemical impedance spectroscopy (EIS), applicable for characterization, modelling and understanding interfacial processes involving the mass and charge transport on surfaces of modified electrodes. The analytical signal generated by the above-mentioned interacting systems can be translated by various electroanalytical methods, specially voltammetry and amperometry, being their main advantages the possibility of analyzing *in situ* smaller amounts of sample, in a shorter time.

Chapter 3 - Carbon-based nanomaterials such as carbon nanotubes (CNT), carbon black and fullerenes are interesting materials for the development of analytical methods. In electrochemistry carbon nanotubes improves the electron transfer and provides an improvement in the analytical sensitivity and selectivity of phenolic compounds determination. Furthermore, the electroanalytical behavior of CNT toward phenolic compounds can be increased using measures in the presence of surfactants. In this way, this work deals with study of a voltammetric method for simultaneous determination of four emerging pollutants Hydroquinone (HQ), Catechol (CT), Guaiacol (GUA) and 4-Nitrophenol (pNF) using a carbon nanotubes paste sensor (37% of CNT and 63% w/w of mineral oil) whose measures were performed in presence of a cationic surfactant (cetylpyridiniumbromide, CPB). The oxidation peaks for the four phenolic compounds were overlapped in the absence of CPB, while a significant increase on oxidation peak current, a shift of potential towards less positive values and a satisfactory separation of oxidation potential in presence of CPB (0.45 mmol L^{-1}) allowed the simultaneous determination. According to relationship of pH and oxidation potential it was observed a Nerstian system where the number of protons equal to the electrons. From chronoamperometric studies and using Cottrell equation the average diffusion coefficients were found to be $5.45x10^{-6}$, $3.35x10^{-5}$, $2.94x10^{-4}$, $3.22x10^{-4}$ cm^2 s^{-1}, respectively for HQ, CT, GUA and pNF. Under optimized condition, i.e. pH 8.0, BR buffer at 0.1 mol L^{-1} and CPB at 0.45 mmol L^{-1} and linear sweep voltammetry at scan rate of 120 mV s^{-1}, limits of detection (LOD) and quantification (LOQ) in μmol L^{-1} were found to be 3.43 and 11.43 (HQ), 1.72 and 5.73 (CT), 1.10 and 3.67 (GUA) and 4.96 and 16.53 (pNF). The method has been applied for simultaneous determination of these phenolic compounds in water samples.

Chapter 4 - The present study focuses on the synthesis of ion imprinted poly(acrylamide-co-EGDMA) using cobalt and cetyltrimethylammonium bromide (CTAB) as double templates and its application for extraction/preconcentration of Co(II) ions from aqueous medium. The preconcentration properties toward cobalt ions for the IIP-CTAB (hierarchically imprinted organic polymer) and NIP-no CTAB (without imprinting process) was evaluated and their characterization was accomplished by SEM, FT-IR, elemental analysis and TGA. IIP-CTAB provided better recognition capacity toward cobalt ions than that one achieved for other metal ions. From breakthrough curve a maximum adsorption capacity of 0.1496 mg g^{-1} for IIP-CTAB was obtained. The analytical feasibility of IIP-CTAB as adsorbent of cobalt ions was evaluated by the on-line coupling of a preconcentration flow injection system to FAAS optimized by means of experimental design methodology. The preconcentration procedure was based on sample loading at pH 8.65 (20.0 mL) through 50.0 mg of IIP-CTAB packed into a mini-column at a fast flow rate of 9.0 mL min^{-1} with further elution with 2.5 mol L^{-1} HNO_3. The linear dynamic range for cobalt was from 5 up to 200.0 μg L^{-1} with limit of quantification (LOQ) of 0.59 μg L^{-1}, preconcentration factor (PF) of 37 and high sample throughput of 22 h^{-1}. The sensitivity of method using IIP-CTAB as adsorbent was 1.75 times higher than one achieved for NIP-no CTAB. Experimental results showed that adsorbent was very stable chemically since it supported more than 200 cycles of preconcentration and desorption using 2.5 mol L^{-1} HNO_3 as eluent. The preconcentration method was used to determine cobalt in water samples.

Chapter 5 - An electrochemical method for phenothiazine (PTZ) determination, based on development of a sensor of carbon nanotube paste/DNA adsorbed on $SiO_2/Al_2O_3/Nb_2O_5$ (designated as SiAlNb), obtained by sol-gel process/mineral oil (60, 30, 10% w/w) was developed. The aim of this work was to evaluate the performance of the carbon nanotube sensor with DNA as catalyst for the voltammetric determination of phenothiazine. The modified electrode showed electrocatalytic activity towards the oxidation of PTZ. Experimental parameters were optimized and the best conditions were obtained: concentration of DNA adsorbed on the silica matrix was 0.5 mg mL^{-1}; in 0.10 M phosphate buffer solution (pH 6.5) using cyclic voltammetry. The sensitivity obtained with the proposed device for the phenothiazine determination, in the concentration range of 2.5 to 12.5 μmol L^{-1}, was about 1.5-fold in relation to carbon nanotube paste. Moreover, from the plot E_{pa} vs. pH, the E_{pa} for PTZ, shifted linearly (linear correlation of 0.9986) to more negative with increasing the solution pH, resulting in a slope of −58.41 mV/pH. This behavior shows, taking into account a Nernstian system, that the electrode reaction for PTZ takes place with a number of protons equal to the electrons. The determination of phenothiazine was carried out by differential pulse voltammetry (DPV) under the best conditions, pulse amplitude of 60 mV and scan rate of 5 mV s^{-1}. The analytical curve built by DPV was ranged from 0.083 up to 16.66 μmol L^{-1} (r = 0.998) with a sensibility 0.8313 μA μmol^{-1} L and precision < 2%. The limits of detection and quantification were found to be 0.023 μmol L^{-1} and 0.072 μmol L^{-1}, respectively. These data demonstrate that the proposed sensor is quite feasible for the determination of phenothiazine group.

In: New Developments in Analytical Chemistry … ISBN: 978-1-63463-427-4
Editor: Breanna Granger

Chapter 1

ANALYTICAL METHODS FOR ORGANIC AND INORGANIC TOXICS DETERMINATION IN COSMETICS PRODUCTS

D. Melucci[1], C. Locatelli[1], M. D'Annunzio[2], M. Rendina[2] and M. Locatelli[2,3,*]

[1]University of Bologna; Department of Chemistry "Giacomo Ciamician", Bologna, Italy
[2]University "G. d'Annunzio" of Chieti-Pescara, Department of Pharmacy, Chieti, Italy
[3]Interuniversity Consortium of Structural and Systems Biology INBB, Roma, Italy

ABSTRACT

The widespread use has made preservatives become a special monitored group in commercial cosmetics, especially in the *leave-on* products due to the prolonged exposure.

Additionally, recently there has been an increase in the demand for preservative-free cosmetics related to undesired adverse effects on human health, in order to minimize the amount in personal care products or

* Corresponding author: Dr. Marcello Locatelli, PhD. Assistant Professor, Analytical and Bioanalytical Chemistry, University "G. d'Annunzio" of Chieti-Pescara; Department of Pharmacy. Interuniversity Consortium of Structural and Systems Biology INBB; Viale Medaglie d'oro 305; 00136 Roma; Italy. E-mail: m.locatelli@unich.it.

produce preservative-free-cosmetics and verify the contents to avoid health risk.

Hence, the simultaneous separation and trace-level determination in personal care products are important for both merchandise reliability and quality assurance.

However, is not possible to exclude that heavy metals can be accumulated in the body over time and repetitive metal-containing product application can lead to significant exposure levels.

In the case of some metallic powders such as nickel (Ni), lead (Pb), and cadmium (Cd), they are unintentionally added to the cosmetics as impurities during manufacturing of an ingredient or product, and represent a minor source of exposure compared to other pathways such as water, food or air.

For these reasons it is necessary a deep knowledge of real toxics (organic and inorganic) content into personal care products and cosmetics using reliable, high-throughput, sensitive and selective analytical assays.

In this chapter, recently developed method for the extraction, separation, and quantitative determination of organic (phthalates and parabens) and inorganic toxic (heavy metals) in cosmetic products are reported.

Keywords: Chemical Analysis, Safety Testing, Spectroscopy, Chromatography, Organic and Inorganic Toxics, Preparative Assay, Analytical Methods

ABBREVIATIONS

HPLC	High performance liquid chromatography
PAEs	phtalate acid esters
MP	Methyl-p-hydroxybenzoate
EP	Ethyl-*p*-hydroxybenzoate
PP	Propyl-*p*-hydroxybenzoate
IPP	Isopropyl-*p*-hydroxybenzoate
BP	Butyl-*p*-hydroxybenzoate
IBP	Isobutyl-*p*-hydroxybenzoate
HP	heptyl-*p*-hydroxybenzoate
BeP	Benzyl-*p*-hydroxybenzoate
DMP	Dimethyl phthalate
DEP	Diethyl phthalate
DAP	Diallyl phthalate

DPP	Di-*n*-propyl phthalate
DBP	Di-*n*-butyl phthalate
DIBP	Di-isobutyl phthalate
BCP	Butyl cyclohexyl phthalate
DNPP	Di-*n*-pentyl phthalate
DIPP	Di-isopentyl phthalate
DCP	Dicyclohexyl phthalate
BBP	Butyl benzyl phthalate
DNHP	Di-*n*-hexyl phthalate
DIHxP	Diisohexyl phthalate
DIHpP	Di-isoheptyl phthalate
BDP	Butyl decyl phthalate
DEHP	Di(2-ethylhexyl) phthalate
DNOP	Di(*n*-octyl) phthalate
DIOP	Di-isooctyl phthalate
ODP	*n*-Octyl *n*-decyl phthalate
DINP	Diisononyl phthalate
DPHP	Di(2-Propyl Heptyl) phthalate
DIDP	Di-isodecyl phthalate
DUP	Di-undecyl phthalate
DIUP	Diisoundecyl phthalate
DTDP	Di-tridecyl phthalate
DITDP	Di-isotridecyl phthalate
B2MEP	Bis(2-methoxyethyl) phthalate
NPIPP	*n*-pentyl-isopentyl phthalate

INTRODUCTION

Recently there is an increasingly demand inherent cosmetics and personal care product composition, in particular related to the possible biological activities of some ingredients [1], and their bioaccumulation into organisms.

Due to the numerous interfering compounds, these analyses remain a very challenging task for manufacturers, and most of the advances in multi-component analysis have been made for the determination of parabens, phthalates, antioxidants [1], or other molecules used in these products.

In this scenario, other kinds of toxics are recently recognized as important target compounds that must be evaluated: heavy metals.

As a result, it is central to minimize the amount of toxics in cosmetics or produce preservative-free products and verify the relevant contents.

Simultaneous separation and trace-level determination in cosmetics are important for both product reliability and quality assurance [2, 3].

Several analytical methods have been reported for the determination of parabens in a variety of matrices, such as phthalates, especially using GC and HPLC coupled with various preconcentration steps such as liquid-liquid extraction (LLE), solid-phase extraction (SPE), and solid-phase microextraction (SPME) [4, 5].

GC is the most used chromatographic method, due to advantages in terms of resolution, separation, low cost, and sensitive and selective detectors such as Mass Spectrometer (MS) [6] that actually can be easily interfaced.

Flow injection analysis (FIA) methods has also used in paraben analysis during the past years, coupled with Capillary Electrophoresis (CE) in order to obtain reproducibility in peak response, retention time, and enhanced sample throughout [5].

Most Phthalate Acid Esters (PAEs) are semi-volatile, thermostable, and non-polar compounds, and for these reasons GC with capillary column is the used configuration, and mass spectrometry (MS) results to be the routine detection technique [7].

Nevertheless, only a few electroanalytical methods have been reported for the determination of parabens [8].

In this chapter, recently developed methods for the extraction, separation and quantitative determination of organic (phthalates and parabens) and inorganic toxics (heavy metals) in cosmetic products are reported, especially related to novel instrument configurations and novel extraction techniques for organic toxics, using reliable, high-throughput, sensitive, and selective assays.

PARABENS

The safety of cosmetic formulation has attracted great attention [4], and an increasing demand to know their composition, in particular due to the possible biological activities of some ingredients, play a central role in quality control (QC) tests.

Preservation is necessary to prevent any alteration related to microbial activities, and to protect the consumers from eventually dangerous degradation products [9-11].

Parabens are widely used as preservatives in cosmetics from microbial and fungal attack, with no perceptible odour or taste; in addition they have neutral pH, and do not produce discoloration and cause hardening.

Their use in cosmetics arises from low toxicity, broad spectrum of activity, inertness, worldwide regulatory acceptance, biodegradability, low cost, and chemical stability related to pH (effective between pH 4.5 and 7.5) and temperature.

Products containing parabens can be safely autoclaved without significant loss of antimicrobial activity, as a result of hydrolysis.

From the chemical point of view, parabens are esters of para-hydroxybenzoic acid, from which the name is derived, and most common parabens used are methylparaben (E218), ethylparaben (E214), propylparaben (E216) and butylparaben, while less common ones include isobutylparaben, isopropylparaben, benzylparaben and their sodium salts.

The general chemical structure of a paraben is shown in Figure 1.

Parabens are often used in combination with other preservatives, since they have a synergistic effect, which allow the use of lower levels while increasing preservative activity. In Table 1 their names, acronyms, structural formula and molecular weight are reported.

Increased carbon number of the ester group correspond to differences in relative activities, increased antimicrobial activity and a decreased water solubility [2, 11, 12].

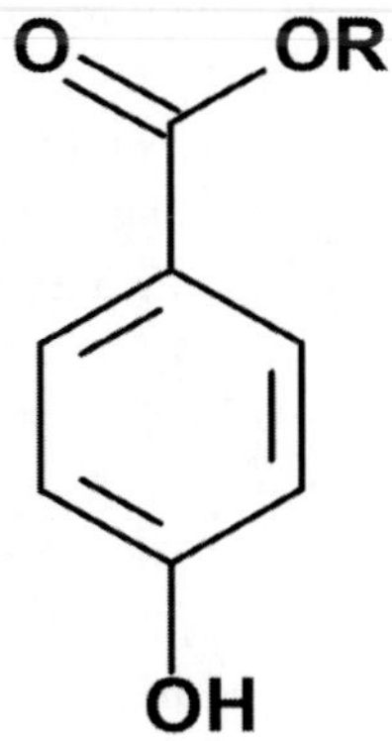

R = alchil group

Figure 1. Parabens general chemical structures.

Table 1. Names, acronyms, structural formula and molecular weight for parabens

Name	Abbr.	E number	Structural formula	MW (g/mol)
Methyl-*p*-hydroxybenzoate	MP	E218	$CH_3(C_6H_4(OH)COO)$	152.15
Ethyl-*p*-hydroxybenzoate	EP	E214	$CH_3CH_2(C_6H_4(OH)COO)$	166.17
Propyl-*p*-hydroxybenzoate	PP	E216	$CH_3CH_2CH_2(C_6H_4(OH)COO)$	180.20
Isopropyl-*p*-hydroxybenzoate	IPP		$CH_3CH_2CH_2(C_6H_4(OH)COO)$	180.20
Butyl-*p*-hydroxybenzoate	BP		$C_4H_9(C_6H_4(OH)COO)$	194.23
Isobutyl-*p*-hydroxybenzoate	IBP		$C_4H_9(C_6H_4(OH)COO)$	194.23
Heptyl-*p*-hydroxybenzoate	HP	E209	$C_7H_{15}(C_6H_4OHCOO)$	236.31
Benzyl-*p*-hydroxybenzoate	BeP		$C_6H_5CH_2(C_6H_4OHCOO)$	228.24

In particular, parabens show inhibitory effects on microbial membrane transport and on the mitochondria function processes [13]. Gruvberger and coworkers analysed 100 dermal moisturizer products by HPLC, reporting that parabens were the most common preservatives in the formulations [14]. Normally, a person employs different cosmetics during a day (e.g. toothpaste, shampoo, and shower gel, body cream, anti-perspirant) and approximately 90% contain one or more parabens [13]. These constituents come in contact with skin and are subjected to a local potential penetration, skin layers accumulation and, in some cases, passage to the general body circulation being transported by the blood to the body tissues [13], as reported in Figure 2. It is important to highlight that the type of cosmetic is irrelevant, while it is fundamental which chemicals are used in the formulations and the application frequencies to the skin [15], related to the fact that toxics can enter as intact esters from topical application of cosmetic products. This is confirmed by the measurement of intact esters in normal human urine and by the demonstration that these molecules can penetrate (via dermal penetration and absorption) into the circulatory system from a single topical cosmetic application.

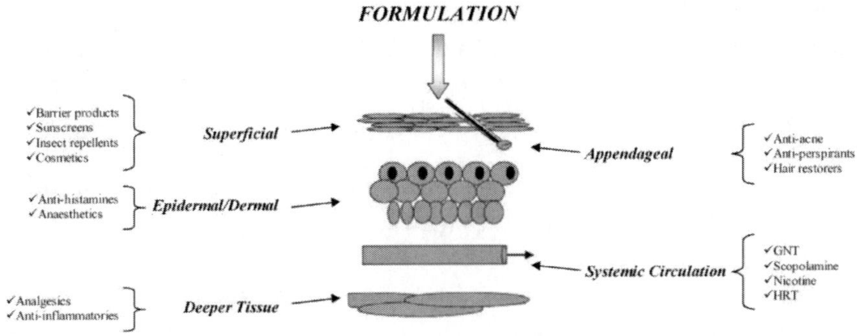

Figure 2. Skin penetration of several cosmetic products.

In recent years, it has been suggested that some cosmetics are a potential source of estrogenic chemicals, and consequently that personal care cosmetics may be associated with the rising incidence of breast cancer in women [14].

Skin esterases convert lipophilic molecules into water-soluble ones through their main metabolite para-hydroxybenzoic acid (which also shows weak estrogenic activity), which is readily excreted into bile, urine, and eliminated, but these enzymes can be saturated in an acute overload situation [13-17]. *In vitro* studies have shown that 30% of applied PP penetrates rat skin and remains intact; after 8 h penetration of some esters can be higher (up to 60% of MP and 40% of EP) and crossing rabbit skin remaining intact [18].

Parabens could be retained in tissues without hydrolysis [13] resulting in long-term accumulation [14]. In addition, recent works have revealed that parabens hydrolysis is slower in human than in rat skin, suggesting that data on animal may underestimate the paraben esters absorbed in human skin.

To this discrepancy, it must be added that formulation variations can influence dermal permeation, related also to the presence of alcohol (acting as esterase inhibitor), thus enhancing dermal penetration [18]. It is interesting to note that the parabens detected at highest levels were MP and PP (now subject to regulatory withdrawal for food uses) [18], revealing the widespread use of these molecules in final products, the ability to be absorbed into tissues, and the inertness with respect to hydrolysis by esterases and subcutaneous fat tissue. On the other hand, BeP was not found and this may be attributed to its less frequent use in products [15]. Most studies reported parabens as not mutagenic, but there are evidences that they may produce chromosomal aberrations, may be related to mammary adenocarcinomas in rats [15], and could produce allergic contact dermatitis [13].

In 2004, the detection of paraben esters in human breast cancer tissue (MP accounted for ca. 60% of the total parabens detected) stimulated international discussion about the safety of using parabens in a wide range of personal care products [18, 14].

Regulatory Status of Parabens

Cosmetics are less stringently tested and receive less regulatory attention compared with other chemicals to which the general population is exposed [18]. Recent research show a high percentage of adverse phenomena, with 26.5% of women and 17.4% of men reporting an event/reaction to cosmetic use. Parabens used in food are more highly regulated and regulatory reviews have resulted in the withdrawal of Acceptable Daily Intake (ADIs).

For example, JECFA 2007 (the Joint Food and Agriculture Organization (FAO) and World Health Organization (WHO) Expert Committee on Food Additives) [19] has recommended *'that in view of the adverse effects in male rats [reproductive toxicity associated with endocrine effects] propylparaben should be excluded from the group ADI for parabens used in food'*, recommending its withdrawal. The same evaluation noted the withdrawal of specification for BP on similar grounds, and involves oral exposures, but it needs to be updated as concerns dermal application.

The European Commissions Scientific Committee on Consumer Products (SCCP) opinion on the safety evaluation of parabens in 2005, requested more information on *'full descriptions of available in vitro percutaneous absorption studies'* and *'a complete dossier with regard to the reproductive and developmental toxicity of PP, IPP, BP and IBP*'.

The European Cosmetic Toiletry and Perfumery Association (COLIPA) provided a submission to this request, but SCCP (2006) [20] evaluated these data as *"not scientifically valid and therefore there still remain inadequacies and deficiencies in the paraben safety dataset, and concerns over paraben endocrine and reproductive toxicity"* [18], and consequently, there is an absolute necessity to regulate repeated applied doses in a long term usage [13, 21]. On 11 July 2013 the EU Cosmetics Directive (76/768/EEC) [22] replaced the EU Cosmetics Regulation (1223/2009) [23], which permits the use of parabens in cosmetics products with a maximum concentration for each one of 0.4% (*w/w*) and total maximum concentration of 0.8% (*w/w*) expressed as para-hydroxybenzoic acid.

PHTHALATES

Esters of phthalic acid (PAEs), commonly phthalates, are used in toys, vinyl flooring and wall covering, detergents, lubricating oils, food packaging, pharmaceuticals, blood bags and tubing, and personal care products, such as nail polish, hair sprays, soaps, and shampoos [11].

They may be present in cosmetics as ingredients or contaminants, and can be related to the manufacturing process, to raw materials used or to the migration of phthalates from packaging when plastic materials are used [10].

As ingredients, some phthalates, particularly those with a low molecular weight, can be used even if they are used primarily as plasticizers in products such as nail polishes (reduce cracking by making them less brittle), and hair sprays (help avoid stiffness by allowing them to form a flexible film on the hair).

Phthalate esters are also used as solvents, perfume fixatives [24], and frequently in cosmetic products DBP, DMP and DEP are present.

Phthalates are dialkyl or alkyl aryl esters of phthalic acid (also called 1,2-benzenedicarboxylic acid, Figure 3), colourless and odourless, that exhibits low evaporation.

Like in the case of parabens, in recent years phthalates have attracted attention due to their suspected carcinogenic and estrogenic properties [4]. For this reason, they were detected in the environment, consumer products, indoor and personal air, indoor dust, and air inside vehicles [25].

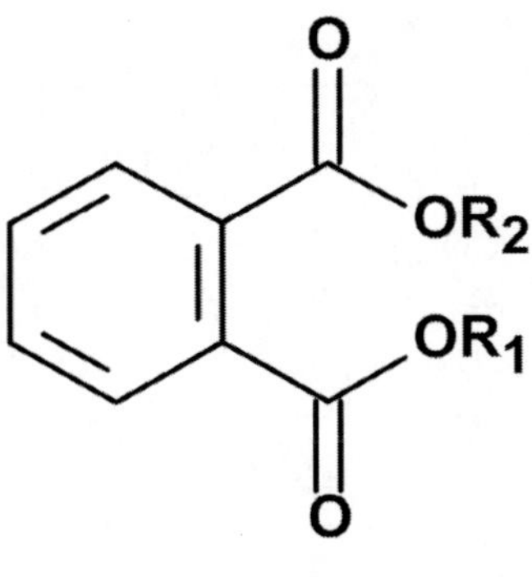

R_1 and R_2 = dialkyl or alkyl aryl group

Figure 3. Chemical structures for phthalates.

The results indicated multiple routes of exposure such as ingestion, inhalation and skin contact, leading to detectable levels in humans [26].

These molecules are so ubiquitous that sample contamination from environment may be a serious problem and can lead to overestimates the exposures [26] during QC analyses. In Table 2 their names, acronyms, structural formula and molecular weight are reported.

Table 2. Names, acronyms, structural formula and molecular weight for phthalates

Name	Abbr.	Structural formula	MW (g/mol)
Dimethyl phthalate	DMP	$C_6H_4(COOCH_3)_2$	194.18
Diethyl phthalate	DEP	$C_6H_4(COOC_2H_5)_2$	222.24
Diallyl phthalate	DAP	$C_6H_4(COOCH_2CH{=}CH_2)_2$	246.26
Di-n-propyl phthalate	DPP	$C_6H_4[COO(CH_2)_2CH_3]_2$	250.29
Di-n-butyl phthalate	DBP	$C_6H_4[COO(CH_2)_3CH_3]_2$	278.34
Diisobutyl phthalate	DIBP	$C_6H_4[COOCH_2CH(CH_3)_2]_2$	278.34
Butyl cyclohexyl phthalate	BCP	$CH_3(CH_2)_3OOCC_6H_4COOC_6H_{11}$	304.38
Di-n-pentyl phthalate	DNPP	$C_6H_4[COO(CH_2)_4CH_3]_2$	306.40
Di-isopentyl phthalate	DIPP	$C_6H_4[COO(CH_2)_2CH(CH_3)_2]_2$	306.40
Dicyclohexyl phthalate	DCP	$C_6H_4[COOC_6H_{11}]_2$	330.42
Butyl benzyl phthalate	BBP	$CH_3(CH_2)_3OOCC_6H_4COOCH_2C_6H_5$	312.36
Di-n-hexyl phthalate	DNHP	$C_6H_4[COO(CH_2)_5CH_3]_2$	334.45
Diisohexyl phthalate	DIHxP	$C_6H_4[COO(CH_2)_3CH(CH_3)_2]_2$	334.45
Diisoheptyl phthalate	DIHpP	$C_6H_4[COO(CH_2)_4CH(CH_3)_2]_2$	362.50
Butyl decyl phthalate	BDP	$CH_3(CH_2)_3OOCC_6H_4COO(CH_2)_9CH_3$	362.50

Name	Abbr.	Structural formula	MW (g/mol)
Di(2-ethylhexyl) phthalate	DEHP, DOP	$C_6H_4[COOCH_2CH(C_2H_5)(CH_2)_3CH_3]_2$	390.56
Di(n-octyl) phthalate	DNOP	$C_6H_4[COO(CH_2)_7CH_3]_2$	390.56
Diisooctyl phthalate	DIOP	$C_6H_4[COO(CH_2)_5CH(CH_3)_2]_2$	390.56
n-Octyl n-decyl phthalate	ODP	$CH_3(CH_2)_7OOCC_6H_4COO(CH_2)_9CH_3$	390.56
Diisononyl phthalate	DINP	$C_6H_4[COO(CH_2)_6CH(CH_3)_2]_2$	418.61
Di(2-Propyl Heptyl) phthalate	DPHP	$C_6H_4[COOCH_2CH(CH_2CH_2CH_3)(CH_2)_4CH_3]_2$	418.61
Diisodecyl phthalate	DIDP	$C_6H_4[COO(CH_2)_7CH(CH_3)_2]_2$	446.67
Diundecyl phthalate	DUP	$C_6H_4[COO(CH_2)_{10}CH_3]_2$	446.67
Diisoundecyl phthalate	DIUP	$C_6H_4[COO(CH_2)_8CH(CH_3)_2]_2$	474.72
Ditridecyl phthalate	DTDP	$C_6H_4[COO(CH_2)_{12}CH_3]_2$	530.82
Diisotridecyl phthalate	DITDP	$C_6H_4[COO(CH_2)_{10}CH(CH_3)_2]_2$	530.82
Bis(2-methoxyethyl) phthalate	B2MEP	$C_6H_4[COO(CH_2)_2OCH_3]_2$	282.29
n-pentyl-isopentyl phthalate	NPIPP	$CH_3(CH_2)_4OOCC_6H_4COO(CH_2)_2CH(CH_3)_2$	306.40

A study commissioned by Greenpeace (2005) [27] reported the presence of phthalates in perfumes sold on the European market, and DEP was found in 35 of the 36 tested products with concentration of 2.23% (w/w) while DMP levels were around 0.3% (w/w).

Phthalate esters share some structural similarities, but data reveal different adverse effects, even if some evidences show that the mechanisms of action and effects of some phthalates are similar enough to be considered to be additive [26].

Studies on the toxicology of phthalates are available since they were introduced into commerce about 75 years ago, but their possible adverse effects in humans are not reported before 25 years and were originally focused on DEHP, which was shown to produce cancer in rodents after high lifetime exposure [26].

Data on phtalates health effects are limited and often inconclusive [28], but several epidemiological studies on animals showed that may be associated with toxicity on male reproduction system, e.g., sperm concentration, even if adverse effects on reproduction (such as reduced fertility) or its development, ovary weight and serum estradiol levels [26] are not reported.

In addition to suspected reproduction toxicity, a British epidemiological study found a link between phthalates detected in indoor air and the development of allergy and asthma in children, but the evidence was limited by imprecise data on levels of exposure due especially to study on rodents [28].

REGULATORY STATUS OF PHTHALATES

Policies and regulations restrict the use of these molecules in various jurisdictions, such as United States and European Union, where phthalates in children's product are regulated [29].

Furthermore, European Union has banned DBP and DEHP in cosmetic and personal care products [25], including in the list of prohibited substances (EU Cosmetics Regulation (1223/2009) [23]):

- ✓ DBP
- ✓ DEHP
- ✓ B2MEP
- ✓ NPIPP
- ✓ DNPP
- ✓ DIPP
- ✓ BBP

Exposure guidance values (estimating daily exposures that are likely to be without an appreciable risk) are also developed for some phthalates, such as DEP, DIBP, DBP, and DEHP.

Particularly, Health Canada established for DEHP and DBP a tolerable daily intake (TDI) of 0.044 mg/kg body weight/die and 0.063 mg/kg body weight/die respectively [30, 31], and more recently the European Food Safety Authority (EFSA, 2005a) [32] established a TDI of 0.048 mg/kg body weight/ die for DEHP related to its reproductive effects and a TDI of 0.01 mg/kg body weight/die for DBP for developmental toxicity in male rodents [25].

Heavy Metals

Environmental heavy metal pollution is considered a worldwide problem, and at least 20 metals are known to be toxic such as lead (Pb), copper (Cu), cadmium (Cd), mercury (Hg), nickel (Ni), silver (Ag), and zinc (Zn).

These metals, released into the environment in relevant quantities, may represent a risk to human health, causing carcinogenesis, mutagenesis, and teratogenesis [33], even if some heavy metals (Fe, Cu, Zn, Co) at moderated concentration are non-toxic [34].

Many cosmetic products contain metals such as ingredients or impurities (due to the release from metallic devices used during their manufacture).

For instance, chromium (Cr) is used in a very small number of products as a colorant, and iron oxides are common colorants in eye shadows, blushes and concealers, while aluminium (Al) compounds are used as colorants in lip-glosses, lipsticks and nail polishes.

Zn and zinc oxide are used in many products, including moisturizers, shampoos, foundations, concealers, and diaper ointments, while titanium dioxide is widely used in sunscreens.

Personal care products such as moisturizing creams, lipsticks, eye cosmetics, shampoos, cleansing milk, henna dye, etc. are commonly used by millions of consumers on a daily basis and represents an important exposure way to these inorganic toxics [35, 36].

It is extensively well-known that Ni and other metals occurring as impurities in cosmetic products might give rise to contact dermatitis in subjects with pre-existing allergy [37].

Toxic metals from cosmetics represent a minor source of exposure compared to other pathways such as water, food or air [36], but represent another possible health risks.

Studies conducted on lipsticks, soaps, creams, hair creams, sunscreen products, eye shadows, henna samples, kohl, eye pencils, eye liners and mascara reveal the presence of lead and other heavy metals [38].

Lip items were considered interesting due to the fact that a significant part of applied product is likely to be ingested, while eye shadows (externally applied) bring to a insignificant skin absorption of Pb [39].

Pb and Cd are described as the most dangerous contaminants in human civilization due to their distribution in the environment as polluting elements, and their toxicity at high levels of exposure is well known.

Furthermore, an increased level of Pb and Cd may cause serious health problems such as both acute and chronic poisoning, pathological changes of organs and diseases related to cardiovascular, kidney, bone, and liver, and may cause cancer owing to excessive accumulation in human body [36].

Pb can cross placenta during pregnancy: this can be related with intrauterine foetal death, premature delivery and low birth weight, just at approximately 10 μg/dL of maternal blood concentration.

Ni is the cause of allergic contact dermatitis (ACD) with an incidence of 20% in females and 1% in males, due to the free-Ni ions released, followed by Co, and Cr [36, 37].

Reactions to Ni have been associated also to Cr, Co, and Pd sensitivity [36], and due to the increased frequency of allergy to this metal [37], it has been proposed that cosmetics must contain less than 5000 ng/g, and levels below 1000 ng/g of other metals (Cr, Co and Ni).

Hg or Hg-derived compounds are most toxic heavy metals [33] that can enter the body through the skin or directly through the mouth [40], damaging brain, heart, kidney, stomach, and intestines even at very low concentration [33].

REGULATORY STATUS OF HEAVY METALS

Cosmetics are subjected to restrictive regulations, such as the European Directive 76/768/EEC, in order to ensure their safety and usefulness [40], and in particular on July 2013 the EU Cosmetics Directive [22] replaced the EU Cosmetics Regulation [23], applied in all EU Member States.

Legislations include information about:

1. Substances banned (these must not be found in the finished product);
2. Substances restricted;
3. Substances subjected to positive listing (used in controlled concentration).

The Directive 76/768/EEC [23] and other revisions banned the use of Cd, Co, Cr, Ni, and Pb (as metallic ions or salts) in the cosmetic formulations [36, 40, 41].

Pb is also listed in Annex II of both Directive 76/768/EEC and EC Reg. 1233/2009 [22, 23], among the substances that must not form part of the composition of cosmetic products [38] in order to respect good manufacturing practice [39].

Imported products could exceed safe limit imposed by European legislation [36] due to the absence of regulations relevant to the export of cosmetics, and for this reason Consume Products Security Commission (CSPC) subscribed an accord with the Chinese authorities in order to control the safety of products, to give exportation certificate [36], and to determine the amount of heavy metals used in cosmetic products [42] by different analytical methods, including Inductively Coupled Plasma-Mass Spectrometry (ICP-MS), Graphite Furnace-Atomic Absorbance Spectroscopy (GF-AAS), Flame AAS, voltammetric detection, Surface Plasma Resonance (SPR), peptide and peptide coated quantum dots, and more recently using fluorescent sensors based on DNAzyme for the detection of a number of metal ions [43].

Sample Pre-Treatment

For cosmetic products analysis, sample pre-treatment is the key step considering the sample complexity, in order to clean up and concentrate the analytes.

Different sample preparation methods have been developed, especially solid phase extraction (SPE), solid phase micro extraction (SPME), supercritical fluid extraction (SFE), dispersive liquid phase micro extraction (DLLME) and shaking and ultrasonic extraction.

SPE and SPME are the main choice for this determination, owing to their simplicity and effectiveness of extraction, coupled to lower solvent consumption for routinely analyses, high selectivity and sensitivity, reducing or eliminating the volume of employed organic solvent.

SPE procedure to extract parabens from cosmetic sample (shampoo, toothpaste, aftershave gel, hair dye) and their determinations using CE was reported in literature [44], while Ting Fei and coworkers [3] proposed a novel SPME fiber preparation method through UV-induced polymerization of poly (ethylene glycol) diacrylate (PEG-DA) for the determination of parabens in cosmetic products, as reported in Figure 4.

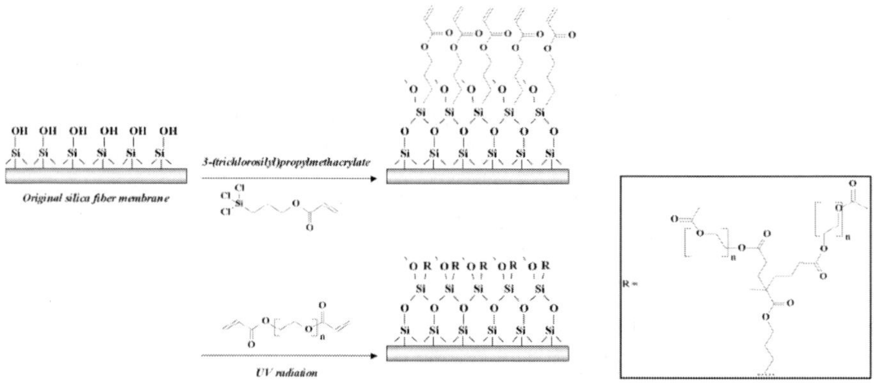

Figure 4. General procedure to UV-induced polymerization of PEG-DA [3].

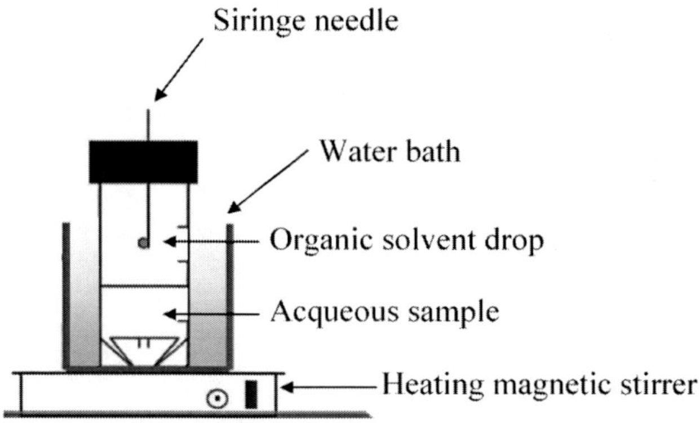

Figure 5. General procedure for SDME extraction protocol.

The PEG-DA polymer coating, covalently bonded to the fibre by a surface modification with 3-(trichlorosilyl) propyl methacrylate (TPM), increase analytical performances by pre-concentration of parabens including MP, EP, PP and BP.

In recent years, Single-Drop MicroExtraction (SDME) has been developed as a valid alternative to SPME, providing extraction in few microliters of an organic solvent, avoiding sample carry-over, fibre degradation, with advantages related to fast, inexpensive and simple equipment for analyses (Figure 5).

SDME method followed by *"in-syringe"* derivatization and GC-MS determination has been developed for analysis of five parabens (MP, EP, IPP, NPP and BP) in cosmetics [6].

Recently Stir Bar Sorptive Extraction method was also used, combined with LC (SBSE-LC) and liquid desorption method for the simultaneous determination of MP, EP, PP, and BP in different cosmetic products [45].

Pressurized Liquid Extraction (PLE), a one-step sample preparation methodology for multi-class preservatives in cosmetics, was used with acetylation followed by GC-MS for the simultaneous determination of two bromine-containing preservatives, seven parabens, IPBC, TCS, and the antioxidant preservatives BHA and BHT in multi-matrix cosmetic samples [46]. Msagati and coworkers [47] developed a simple and direct method based on Hollow Fiber Supported Liquid Membrane (HFSLM, Figure 6) extraction and liquid chromatography with UV detector for the analysis of parabens in cosmetic products (MP, EP, PP, BP and IBP).

In HFSLM extraction the sample, in contact with few microliter of organic solvent immobilized onto hollow fibers membrane pores, is selectively extracted and enriched in the acceptor solution inside the lumen of the hollow fibers [47].

Márquez-Sillero and coworkers [48] developed carbon nanotubes for the solid phase extraction of parabens in cosmetic products.

These nanometric materials have a high hydrophobic surface area and exhibit strong interaction capabilities for various compounds. As a result, Multi-Walled Carbon NanoTubes (MWCNTs) have been successfully used as sorbent material for the determination of organic compounds in different matrices [48].

Matrix Solid-Phase Dispersion (MSPD, Figure 7) was firstly introduced as a simple, efficient and inexpensive method of sample preparation, and recently it was applied to different preservatives such as bromine-containing preservatives, parabens, IPBC, TCS, and the antioxidants BHA and BHT, in a broad spectrum of cosmetic samples, including both rinse-off and leave-on products [5].

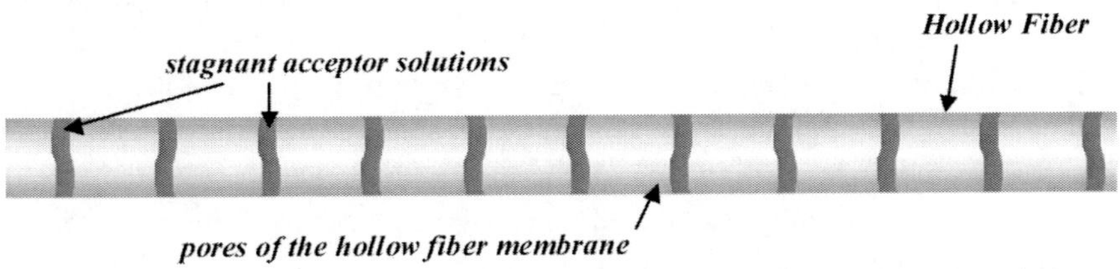

Figure 6. General procedure for HFSLM extraction protocol.

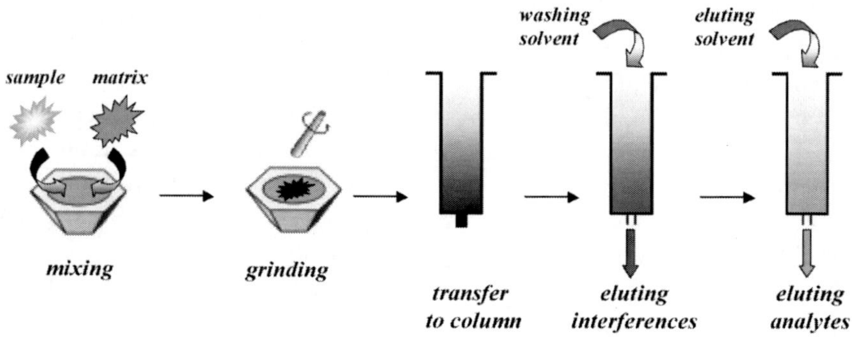

Figure 7. General procedure for MSPD extraction protocol.

Supercritical Fluid Extraction (SFE) is a relatively recent green extraction assay suitable as a valuable technique for the isolation of solutes from solid samples [9], especially in the case of essential oils extraction from raw plant material. It offers many advantages compared to conventional extraction:

- ✓ Less consumption of hazardous organic solvents
- ✓ Reduction in laboratory labour
- ✓ Higher sample throughput
- ✓ Gentler environment for the analytes (oxygen-free and relatively low temperature)
- ✓ Production of cleaner extracts

SFE combined with HPLC–MS to determine trace preservatives and antioxidants in cosmetic products was reported in literature [9].

Scalia and Games describe the development of an SFE procedure, performed with a commercially available system, for the rapid and efficient purification of cosmetic matrices before paraben analyses by HPLC [49].

Another work describes the development of an improved SFE coupled with Capillary Zone Electrophoresis (CZE) for the determination of MP, EP, PP, BP (isolated with co-solvent-induced SFE system) in creams [50].

Today, the miniaturization of conventional sample preparation techniques is one of the main tasks in analytical chemistry (evidenced by the number of articles published on this subject in the last two decades) and microextraction techniques have been gaining a growing interest.

In recent years Dispersive Liquid–Liquid Phase MicroExtraction (DLLME) and Emulsification Liquid Phase MicroExtraction (ELPME) are considered significant methods for miniaturized sample preparation [51], owing to advantages such as:

- ✓ Use of microliter of extraction solvent (environmentally friendly)
- ✓ Short extraction time (rapidly achieving the equilibrium state)
- ✓ High enrichment factor (high phase ratio)

In this technique, centrifugation is considered to be the time-consuming step and a real bottleneck in the automation, and up to now its application has been necessary in the majority of DLLME and ELPME procedures (Figure 8).

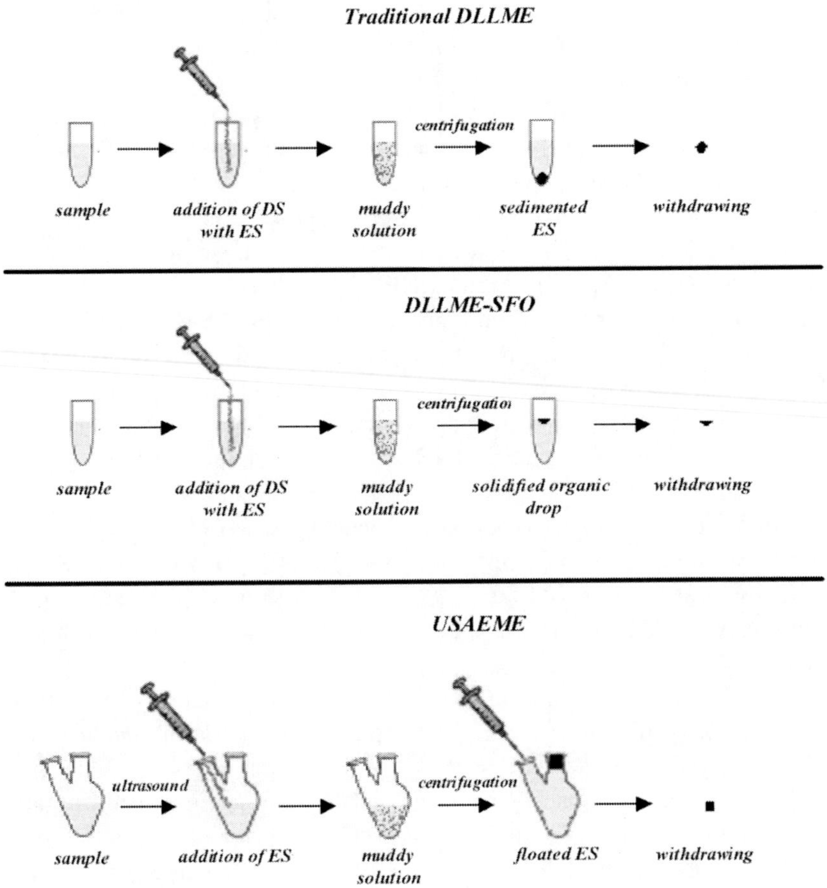

Figure 8. General procedure for different DLLME extraction protocols.

To avoid this problem, Ebrahimpour and coworkers [51] developed a novel *on-line* emulsification microextraction coupled to HPLC, without using centrifugation as separation phase.

Nowadays only few analytical methods for the determination of phthalates have been proposed in the literature, and HPLC-UV/Vis is the most commonly described in cosmetics [52-54].

If several papers are mainly focused on phthalates quantification in cosmetics (DEP, DBP, DEHP and BBP) [24, 52, 54-56], some papers also discuss the determination of less used phthalates [1, 7, 10, 53, 57].

Recently a method was published for the determination of 12 phthalates in cosmetics [10] using a GC-MS system with electron impact ionisation mode (EI), and using the analytical approach described in the ISO 12787 international standard "Cosmetics–Analytical methods–Validation criteria for analytical results using chromatographic techniques" [58].

This method was applied to cosmetic products considered *"ready for analytical injection"* (perfumes), and more complex cosmetic samples (shampoo, cream, shower gel, body milk).

De Orsi and coworkers reported a simple and rapid analytical method for the determination of phthalates [52], based on an ultrasonic extraction with ethanol–water (90:10, *v/v*) followed by HPLC-UV detection.

252 cosmetic and personal care products including 98 baby care products obtained from the Canadian market in year 2007 [56] were analysed, after extraction with different organic solvents, followed by GC–MS analysis.

The main objective was to conduct a national survey of phthalates levels in cosmetic and personal care products to document the potential exposure related to the use.

Chingin and coworkers reported a novel procedure based on Extractive ElectroSpray Ionization Mass Spectrometry (EESI-MS) for fast detection and identification of DEP in perfumes without any sample pre-treatment [55] within a few seconds, providing a semi-quantitative information on the DEP content.

Due to the low concentration of PAEs in real samples and complex matrix interference, it is often essential to perform the enrichment of analytes before GC-MS analysis, and several procedures have been developed using the reduction of organic suspension followed by sonication/ centrifugation/ filtration prior to analysis.

Magnetic Solid-Phase Extraction (MSPE) shows advantages in sample preparation, such as easy operation and short extraction time.

Carbon NanoTubes (CNTs) have strong adsorption capability towards hydrophobic organic compounds, and a rapid, sensitive and cost-effective method for the determination of PAEs was established by coupling the MSPE technique with GC-MS (MSPE-GC-MS) [7] for the analyses in beverage, environmental water and perfume samples.

Polymer Monolith MicroExtraction (PMME) has been recently introduced in order to reduce the use of toxic solvents, and for its high enrichment factors (small extracting solvent ratio with respect to the whole sample volume), integration of extraction, and injection in a simple device that can be easily automated [54] in a sort of *"in-tube"* procedure.

Su and coworkers exploit the potential of PMME using a poly(MAA-EGDMA) monolithic capillary column for the extraction of organic pollutants from cosmetics, and four phthalate esters (DMP, DBP, DCHP, and DEHP) have been selected as model compounds bringing to valuable recovery values [54].

In a recent work, six different micro-scale methods for extraction are used to extract ten phthalate esters (C1 to C8), including Liquid-Liquid Extraction (LLE), Ultrasonic Assisted Extraction (UAE), Microwave-Assisted Extraction (MAE), Dispersive Liquid-Liquid MicroExtraction (DLLME), Dispersive Liquid-Liquid MicroExtraction Solidification of Floating Organic droplet (DLLME-SFO) and Cloud Point Extraction (CPE) [57].

UltraSound Assisted Emulsification MicroExtraction (USAEME, Figure 8) is an effective technique among the microextraction methods due to a micro volume of water-immiscible extraction solvent dispersed into aqueous-solution samples by ultrasound-assisted emulsification without using any dispersive solvent [53], with a very efficient and fast analyte extraction.

However, USAEME requires high-density extraction solvents, such as dichloromethane, carbon tetrachloride, chloroform, tetrachloroethylene and trichloroethane, toxic and environmentally unfriendly solvents [53].

To avoid this problem, Liquid-Phase MicroExtraction based on Solidification of Floating Organic droplet (LPME-SFO), was successfully proposed.

In LPME-SFO a micro drop of extraction solvent with low density and proper melting point near room temperature is delivered to the sample surface, the extraction solvent is then solidified at low temperatures, and easily collected.

In literature, only two studies report the simultaneous determination of both phthalates and parabens [1, 4].

Shen and coworkers [4] reported comparative studies on the simultaneous determination of seven phthalates (DEP, DPP, DBP, BBP, DCHP, DEHP, and DOP), and four parabens (MPB, EPB, PPB, and BPB), in cosmetic products by HPLC-DAD and GC-EI-MS with selected-ion monitoring.

Sanchez-Prado and coworkers described a fast, simple and reliable multi-component analytical method applicable for quality control with a high throughput level, developed to analyse commercial brands of perfumes [1].

Chromatographic Assays

Detection of parabens can be achieved by various analytical methods including HPLC, GC, CE, MS, and some other techniques, like Flow Injection Analysis (FIA).

Mass spectrometry is used for trace-level analysis, but it still requires considerable time and effort, and it is an expensive instrumentation.

Therefore, HPLC is more effective and it is an ideal technique for residual analysis of these components.

The seventh Directive 96/45/EC [59] of the European Commission reports a method for the analysis of cosmetic products, which discloses the determination of parabens by HPLC, but recently a Commission Decision has been published, implementing Council Directive 96/23/EC concerning the analytical methods performance and results interpretation [60].

Borremans and coworkers [12] reported the simultaneous analysis of preservatives in cosmetic products according to the method adopted by the European Commission [12].

Cha and coworkers evaluated an HPLC method for the trace-level simultaneous determination of 19 preservatives (including 6 parabens) commonly found in cosmetics with an easy sample extraction procedure [2].

Fast and sensitive HPLC method for analysis of cosmetic creams for a mixture of hydroquinone (HQ), phenol (PO) and six preservatives (MP, EP, PP, BP, IPP, IBP) [61] has recently been published.

Scalia and Games [49] developed a simple SFE procedure for the rapid and efficient purification of the complex cosmetic matrices before the analyses of parabens by HPLC.

An optimisation of this work allowed for simultaneous quantification of HQ, PO, benzoic acid (BA), sorbic acid (SA), MP, EP, PP, BP, and it was applied to the analysis of 10 different cream samples [62].

Other works reported HPLC analyses for parabens [3, 47, 48, 51], and currently this instrumentation has been widely used by the industry.

High Temperature Liquid Chromatography (HTLC) refers to HPLC separations at elevated temperatures using organic solvent–water mixtures as the mobile phase, and reduced organic solvent, while SuB-critical Water Chromatography (SBWC) is performed at elevated temperatures using pure water as the unique mobile-phase. Unfortunately, the majority of the SBWC research up to date is still at the academic level, and quantitative analysis of real samples has rarely been investigated [63].

Ultra Performance Liquid Chromatography (UPLC) method for determination of 21 preservatives in 8.5 min in cosmetic with ultrasonication extraction [64], and a method for the fast determination of only MP by UPLC [65] were also reported.

Mincea and coworkers developed a rapid UPLC method for the analysis of MP, EP, PP, and BP in 2.5 min, bringing to an assay suitable for routine quality control of cosmetics [66].

In recent years, other separation assays such as Capillary Electrophoresis (CE) and Micellar ElectroKinetic Chromatography (MEKC) have been attractive techniques due to their high efficiency, low reagent consumption and high analysis speed, even if discontinuous sample injection, limited analysis throughput, low sensitivity, and large matrix effects still remain their main limits.

"On-line" sample-stacking methods were developed for sensitivity enhancement. They are divided into two injection modes [67]:

1 Electrokinetic injection
2 Hydrodynamic injection

Uysal and Güray described a validated CE method for the determination of MP, EP, PP, and BP [44] like Wang and Chang in cosmetic products [50].

Comparison between HPLC and CZE for the determination of MP, EP, PP, and BP in cosmetic shampoos with an easy sample extraction using acidified diethyl ether prior to injection was also reported [68].

An alternative assay for the analysis of trace parabens based on MEKC, with improved resolution and better sensitivity by combining large volume sample stacking (LVSS) technique, was applied in parabens for low levels determinations [69, 70]. García Jiménez and coworkers carried out parabens determination with a hybrid FIA-HPLC system with ultra-short monolithic column, resulting in a versatile and efficient system used for the determination of four preservatives (MP, EP, PP, and BP) [71].

FIA present advantages such as simplicity, low cost, versatility, speed, and can include separation steps mainly based on extraction, ion exchange or adsorption to improve selectivity, especially if based on solid phase supports placed in micro columns before the flow cell or directly in the flow cell.

The combination of FIA, Sequential Injection Analysis (SIA) or MultiSyringe Flow Injection Analysis (MSFIA) with chromatography results in the so-called hybrid FIA-HPLC, Sequential Injection Chromatography (SIC) or MultiSyringe liquid Chromatography (MSC), extending the possibilities of flow techniques [71].

An application is based on an ultra-short monolithic column and chemiluminescent detection in a FIA-type instrument for the determination of four paraben mixtures (MP, EP, PP, and BP) [72] and for antioxidants, preservatives and sweetener additives in cosmetics [73].

Several papers report procedures for the simultaneous separation and determination of antioxidants and/or preservatives in cosmetic products by SPME with GC–MS [6, 46, 74, 75] in hyphenated configuration [76-78].

Direct-Analysis-in-Real-Time Mass Spectrometry (DART-MS) is a relatively new analytical technique allowing the measurement of solid, liquid and gaseous samples directly without further pre-treatment, and coupled to mass spectrometer (like a time-of-flight). Spectra with exact masses within a few seconds can be obtained, allowing the unambiguous identification of the analytes [79].

Electrochemical Assays

The choice of the adequate technique often depends on several factors: for example, instrumental availability and especially the purchase and maintenance cost, number of samples to be analysed and so on.

Only in recent years electroanalytical methods have aroused a growing interest for the determination of both organic and inorganic species – even if the last in a very limited manner – in cosmetic matrices. This is certainly due to the low cost of purchase and maintenance of the equipment, together with the possibility of carrying out multicomponent analysis in complex matrices by means of such technique.

Electroanalytical methods are characterised by some fundamental lines and survey fields: a) ion-selective electrodes and modified electrodes, b) voltammetric and c) potentiometric stripping analysis, d) cyclic voltammetry and, finally, e) solid electrodes.

a. Ion-selective electrodes and modified electrodes

The technique is generally addressed to major elements like alkaline, earth-alkaline metals, halogen anions and organic acids [80], since measurements employing ion-selective electrodes are seldom possible at concentrations below mg/L concentration level.

Two interesting applications of ion-selective electrodes are discussed in the works published by Wang, regarding the determination of sulphur in anti-acne cosmetic products [81] and of fluorine and sodium monofluorophoshate in oral hygiene products [82].

In any case, the weakness of the ion-sensitive electrodes concerns precisely the unsatisfactory limit of detections, as previously pointed out. For this reason, the natural evolution of the ion-sensitive/selective electrodes was the development of chemically modified electrodes. In the last decade, this field of research has shown a significant impulse with the publication of many works.

As regards specifically the cosmetic matrices, such electrodes are addressed to different elements and compounds of interest. Still in the cosmetic field, these modified electrodes have been widely used for the determination of organic species, while the application to inorganic species has been much more limited. Regarding the latter species, the research field in relation to toxic metals has been addressed mainly to the determination of mercury [83-87]. Only one work on lead is present in literature, and it reports its determination in cosmetic-hair dye formulations [88].

A work which also deserves to be highlighted concerns the determination of persulfate in hair bleaching boosters by flow injection amperometric determination using a prussian blue film-modified electrode [89].

Still with regard to the inorganic species, it is appropriate to highlight some interesting works concerning the possibility of determining the sulphur in anti-acne creams, although the detection limit is not to be very satisfactory (0.325 mg/L) [90], and hydrogen peroxide [91, 92].

As for the organic species, the interest is addressed to some specific compounds such as methylparaben [93], hydroquinone [94], 8-hydroxyquinoline [95, 96], surfactants [97], benzoyl peroxide [98] and ellagic acid [99, 100].

b. Voltammetric and
c. potentiometric analysis

In fact, only voltammetric and potentiometric determinations, in all their different variants, have great importance. They are addressed to both inorganic and organic species with redox functional groups, and show important advantages: very good accuracy and especially extremely low limits of detection, of the order of μg/L and ng/L. In addition, these techniques allow to perform multicomponent analysis and also to obtain, in addition to the quantitative data, also the qualitative data.

In any case, in the specific field of determinations in cosmetic matrices, such techniques have been employed to quantify different compounds, but all typologically referable to a limited number of categories.

Very interesting are two works on parabens [101, 102], which are usually determined by chromatography.

Other compounds under investigation result to be the disinfectants, employed either in cosmetic products [103] or as anti-plaque and anti-caries agents in toothpaste [104].

Interesting are also the analytical procedures for the determination of sunscreen agents [105, 106] and azo dyes [107] or cosmetic synthetic colouring matters [108], widely employed in cosmetic products.

As regards the determination of toxic metals in cosmetic matrices, the literature is very limited.

Worthy of interest may be two works, the first concerning the determination of cadmium, copper, lead and zinc in essential oils by derivative potentiometric stripping analysis [109], the second relevant to selenium quantification in seed oils, employed in cosmetic, but also in pharmaceutical and alimentary products, by means of cathodic stripping potentiometry [110].

d. Cyclic voltammetry

Cyclic voltammetry is an important technique because, in addition to the ability to determine the reversibility of the electrodic process of the redox species, it allows to carry out quantitative determinations.

In this field, significantly important are the works dealing with parabens [102, 111] (even if such technique has been employed also to investigate the electrodic behaviour of other compounds involved in cosmetic matrices, such as the sulphur in anti-acne products) [90], hexacyanoferrate(III) as sunscreen in commercial cosmetic products [112], phenols and α-tocopherol in pharmaceuticals and cosmetics [113] and hydrogen peroxide [114].

e. Solid electrodes

As regards the use of solid electrodes for the determination of toxic metals in cosmetic matrices, at the moment there are no studies in the literature. The interest is aimed rather to the use of solid electrodes as electrochemical sensors for chromatographic measurements of parabens [115] and Lithol Rubine B [116]. For the latter, an azo dyes cosmetic product, the literature reports also an interesting work on its determination using a glassy carbon electrode as the working electrode [107].

Moreover, glassy carbon electrode was employed for the determination of two anti-oxidants, butylated hydroxyanisol (BHA) and tert-butyl-hydroxyquinone (TBHQ) in cosmetic sun tanning lotions [117], and of aminophylline in thigh creams and in the treatment of asthma products [118].

Finally, in this part of the discussion it is important to highlight a very interesting work published by Chang and Chang relevant to the determination of ascorbic acid (AA) and two ascorbic acid derivatives – magnesium ascorbyl phosphate (MAP) and ascorbylpalmitate (AP) – in cosmetics [119]. These compounds are extremely important, considering that they are believed to affect the processes of skin aging and that they are widely used in skin care products.

CONCLUSION

Parabens and phthalates can be considered as ubiquitous pollutants due to their widespread use not only as preservatives but also as plasticizers (substances added to plastics materials in order to increase their flexibility, transparency, durability and longevity).

Their use is also related to the production of enteric coatings of pharmaceutical tablets and nutritional supplements to viscosity control agents, gelling agents, film formers, stabilizers, dispersants, lubricants, binders, emulsifying agents, and suspending agents. End-applications include adhesives and glues, electronics, agricultural adjuvant, building materials, personal-care products, medical devices, detergents and surfactants, packaging, children's toys, modelling clay, waxes, paints, printing inks and coatings, pharmaceuticals, food products, and textiles.

Due to these large human exposure sources, their determination in cosmetic products is very important, such as the determinations of heavy metals.

These classes of compounds (organic and inorganic) can produce adverse effects and severe diseases in human.

For these reasons it is necessary a deep knowledge of real toxics (organic and inorganic) content into personal care products and cosmetics using reliable, high-throughput, sensitive and selective analytical assays.

As highlighted in this review, to obtain reliable, accurate and reproducible data several recently developed method for the extraction, separation, and quantitative determination of organic (phthalates and parabens) and inorganic toxics (heavy metals) in cosmetic products are reported.

REFERENCES

[1] Sanchez-Prado, L., Llompart, M., Lamas, J. P., Garcia-Jares, C., and Lores, M. Multicomponent analytical methodology to control phthalates, synthetic musks, fragrance allergens and preservatives in perfumes. *Talanta*. 85(1), 370-379 (2011).

[2] Cha, N.-R., Lee, J.-K., Jeong, H.-J., Cho, J.-C., Kim, M.-J., and Lee, S.-Y. Determination of 19 preservatives in various matrices by high-performance liquid chromatography. *Anal. Lett.* 45, 2148-2160 (2012).

[3] Fei, T., Li, H., Ding, M., Ito, M., and Lin, J.-M. Determination of parabens in cosmetic products by solid-phase microextraction of poly (ethylene glycol) diacrylate thin film on fibers and ultra high-speed liquid chromatography with diode array detector. *J. Sep. Sci.* 34(13), 1599-1606 (2011).

[4] Shen, H.-Y., Jiang, H.-L., Mao, H.-L., Pan, G., Zhou, L., and Cao, Y.-F. Simultaneous determination of seven phthalates and four parabens in cosmetic products using HPLC-DAD and GC-MS methods. *J. Sep. Sci.* 30(1), 48-54 (2007).

[5] Han, F., He, Y.-Z. and Yu, C.-Z. On-line pretreatment and determination of parabens in cosmetic products by combination of flow injection analysis, solid-phase extraction and micellar electrokinetic chromatography. *Talanta*. 74(5), 1371-1377 (2008).

[6] Saraji, M. and Mirmahdieh, S. Single-drop microextraction followed by in-syringe derivatization and GC-MS detection for the determination of

parabens in water and cosmetic products. *J. Sep. Sci.* 32(7), 988-995 (2009).

[7] Luo, Y.-B., Yu, Q.-W., Yuan, B.-F., and Feng, Y.-Q. Fast microextraction of phthalate acid esters from beverage, environmental water and perfume samples by magnetic multi-walled carbon nanotubes. *Talanta* 90, 123-131 (2012).

[8] Martins, I., Carreira, F. C., Canaes, L. S., de Souza Campos Junior, F. A., da Silva Cruz, L. M., and Rath, S. Determination of parabens in shampoo using high performance liquid chromatography with amperometric detection on a boron-doped diamond electrode. *Talanta.* 85(1), 1-7 (2011).

[9] Lee, M.-R., Lin, C.-Y., Li, Z.-G., and Tsai, T.-G. Simultaneous analysis of antioxidants and preservatives in cosmetics by supercritical fluid extraction combined with liquid chromatography–mass spectrometry. *J. Chromatogr. A.* 1120(1-2), 244-251 (2006).

[10] Gimeno, P., Maggio, A.-F., Bousquet, C., Quoirez, A., Civade, C., and Bonnet, P.-A. Analytical method for the identification and assay of 12 phthalates in cosmetic products: Application of the ISO 12787 international standard "Cosmetics–Analytical methods–Validation criteria for analytical results using chromatographic techniques". *J. Chromatogr. A.* 1253, 144-153 (2012).

[11] http://www.fda.gov.

[12] Borremans, M., Van Loco, J., Roos, P., and Goeyens, L. Validation of HPLC Analysis of 2-Phenoxyethanol, 1-Phenoxypropan-2-ol, Methyl, Ethyl, Propyl, Butyl and Benzyl 4-Hydroxybenzoate (Parabens) in Cosmetic Products, with Emphasis on Decision Limit and Detection Capability. *Chromatographia.* 59(1), 47-53 (2004).

[13] El Hussein, S., Muret, P., Berard, M., Makki, S., and Humbert, P. Assessment of principal parabens used in cosmetics after their passage through human epidermis-dermis layers (ex-vivo study). *Exper. Dermatol.* 16(10): 830-836 (2007).

[14] Harvey, P. W. and Darbre, P. Endocrine disrupters and human health: could oestrogenic chemicals in body care cosmetics adversely affect breast cancer incidence in women? *J. Appl. Toxicol.* 24(3), 167-176 (2004).

[15] Darbre, P. D., Aljarrah, A., Miller, W. R., Coldham, N. G., Sauer, M. J., and Pope, G. S. Concentrations of Parabens in Human Breast Tumours. *J. Appl. Toxicol.* 24(1), 5-13 (2004).

[16] Janjua, N. R., Mortensen, G. K., Andersson, A., Kongshoj, B., Skakkebaek, N. E., and Wulf, H. C. Systemic uptake of diethyl phthalate, dibutyl phthalate, and butyl paraben following whole-body topical application and reproductive and thyroid hormone levels in humans. *Environ. Sci. Technol.* 41(15), 5564-5570 (2007).

[17] Boberg, J., Taxvig, C., Christiansen, S., and Hass, U. Possible endocrine disrupting effects of parabens and their metabolites. *Repr. Toxicol.* 30 (2), 301-312 (2010).

[18] Darbre, P. D. and Harvey, P. W. Paraben esters: review of recent studies of endocrine toxicity, absorption, esterase and human exposure, and discussion of potential human health risks. *J. Appl. Toxicol.* 28(5), 561-578 (2008).

[19] JECFA. 2007. Evaluation of certain food additives and contaminants. Sixty-seventh report of the joint FAO/WHO Expert committee on Food Additives. *WHO Technical Report Series 940*. World Health Organization.

[20] SCCP. 2006. Scientific Committee on Consumer Products. Opinion on Parabens. COLIPA No. P82. SCCP/1017/06. European Commission, Health and Consumer Protection Directorate-General. CCP. 2005.

[21] SCCP. 2005. Scientific Committee on Consumer Products. Extended Opinion on parabens, underarm cosmetics and breast cancer. SCCP/0874/05. European Commission, Health and Consumer Protection Directorate-General.

[22] European Union (EU) Cosmetics Directive (76/768/EEC).

[23] EU Regulation (EC) No 1223/2009.

[24] Chen, H., Wang, C., Wang, X., Hao, N., and Liu, J. Determination of phthalate esters in cosmetics by gas chromatography with flame ionization detection and mass spectrometric detection. *Int. J. Cosm. Sci.* 27(4), 205-210 (2005).

[25] De Coster, S. and van Larebeke, N. Endocrine-Disrupting Chemicals: Associated Disorders and Mechanisms of Action. *J. Environ. Public Health.* 2012 Article ID 713696, 52 pages http://dx.doi.org/10.1155/2012/713696 (2012).

[26] Kamrin, M. A. Phthalate Risks, Phthalate Regulation, and Public Health: A Review *J. Toxicol. Environ. Health, Part B* 12(2), 157-174 (2009).

[27] Peters, R. J. B. *TNO-report*, R and I-A R 2005/011 Phthalates and Artificial Musks in Perfumes.

[28] Koniecki, D., Wang, R., Moody, R. P., and Zhu, J. Phthalates in cosmetic and personal care products: Concentrations and possible dermal exposure. *Environm. Res.* 111(3), 329-336 (2011).

[29] EC [Health Canada], 2009b Proposed Phthalates Regulations for Soft Vinyl Children's Toys and Child Care Articles. http://hc-sc.gc.ca/cps-spc/legislation/consultation/phthalates-phthalates-eng.php.

[30] EC/EC [Health Canada/Environment Canada], 1994a. Dibutyl phthalate. *Canadian Environmental Protection Act, Priority Substances List Assessment.* Ottawa (ON): Minister of Supply and Services of Canada, 34 pp.

[31] EC/EC [Health Canada/Environment Canada] Bis(2-ethylhexyl) phthalate. *Canadian Environmental Protection Act, Priority Substances List Assessment.* Minister of Supply and Services of Canada, Ottawa (ON) (1994) 44 pp.

[32] European Food Safety Authority. 2005a. Opinion of the Scientific Panel on Food Additives, Flavourings, Processing Aids and Materials in Contact with Food on a request from the Commission related to DINP for use in food contact materials (July 2005).

[33] Wang, Y., Yang, H., Pschenitza, M., Niessner, R., Li, Y., Knopp, D., and Deng, A. Highly sensitive and specific determination of mercury(II) in water, food and cosmetic samples with an ELISA based on a novel monoclonal antibody. *Anal. Bioanal. Chem.* 403, 2519-2528 (2012).

[34] Rojas, E., Herrera, L. A., Poirier, L. A., and Ostrosky-Wegman, P. Are metals dietary carcinogens? *Mutat. Res.* 443, 157-181 (1999).

[35] Gondal, M. A., Seddigi, Z. S., Nasr, M. M., and Gondal, B. Spectroscopic detection of health hazardous contaminants in lipstick using Laser Induced Breakdown Spectroscopy. *J. Hazard. Mater.* 175(1-3), 726-732 (2010).

[36] Volpe, M. G., Nazzaro, M., Coppola, R., Rapuano, F., and Aquino, R. P. Determination and assessments of selected heavy metals in eye shadow cosmetics from China, Italy, and US. *Microchem. J.* 101, 65-69 (2012).

[37] Bocca, B., Forte, G., Petrucci, F., and Cristaudo, A. Levels of nickel and other potentially allergenic metals in Ni-tested commercial body creams. *J. Pharm. Biomed. Anal.* 44, 1197-1202 (2007).

[38] Piccinini, P., Piecha, M. and Fortaner Torrent, S. European survey on the content of lead in lip products. *J. Pharm. Biomed. Anal.* 76, 225-233 (2013).

[39] Al-Saleh, I., Al-Enazi, S. and Shinwari, N. Assessment of lead in cosmetic products. *Reg. Toxicol. Pharmacol.* 54(2), 105-113 (2009).

[40] Jia, X., Han, Y., Wei, C., Duan, T., and Chen, H. Speciation of mercury in liquid cosmetic samples by ionic liquid based dispersive liquid–liquid microextraction combined with high-performance liquid chromatography-inductively coupled plasma mass spectrometry. *J. Anal. At. Spectrom.* 26, 1380-1386 (2011).

[41] Lavilla, I., Cabaleiro, N., Costas, M., de la Calle, I., and Bendicho, C. Ultrasound-assisted emulsification of cosmetic samples prior to elemental analysis by different atomic spectrometric techniques. *Talanta.* 80(1), 109-116 (2009).

[42] Lee, S.-M., Jeong, H.-J. and Seop Chang, I. R and D. Simultaneous determination of heavy metals in cosmetic products. *J. Cosmet. Sci.* 59, 441-448 (2008).

[43] Zuo, P., Yin, B.-C. and Ye, B.-C. DNAzyme-based microarray for highly sensitive determination of metal ions. *Biosens. Bioelectron.* 25(4), 935-939 (2009).

[44] Uysal, U. D. and Güray, T. Determination of Parabens in Pharmaceutical and Cosmetic Products by Capillary Electrophoresis. *J. Anal. Chem.* 63 (10), 982-986 (2008).

[45] Melo, L. P. and Queiroz, M. E. C.; Simultaneous analysis of parabens in cosmetic products by stir bar sorptive extraction and liquid chromatography. *J. Sep. Sci.* 33, 1849-1855 (2010).

[46] Sanchez-Prado, L., Lamas, J. P., Lores, M., Garcia-Jares, C., and Llompart, M. Simultaneous in-cell derivatization pressurized liquid extraction for the determination of multiclass preservatives in leave-On cosmetics. *Anal. Chem.* 82(22), 9384-9392 (2010).

[47] Msagati, T. A. M., Barri, T., Larsson, N., and Jönsson, J. Å. Analysis and quantification of parabens in cosmetic products by utilizing hollow fibre-supported liquid membrane and high performance liquid chromatography with ultraviolet detection. *Int. J. Cosm. Sci.* 30(4), 297-307 (2008).

[48] Márquez-Sillero, I., Aguilera-Herrador, E., Cárdenas, S., and Valcárcel, M. Determination of parabens in cosmetic products using multi-walled carbon nanotubes as solid phase extraction sorbent and corona-charged aerosol detection system. *J. Chromatogr. A.* 1217(1), 1-6 (2010).

[49] Scalia, S. and Games, D. E. Determination of parabens in cosmetic products by supercritical fluid extraction and high-performance liquid chromatography. *Analyst.* 117, 839-841 (1992).

[50] Wang, S. P. and Chang, C. L. Determination of parabens in cosmetic products by supercritical fluid extraction and capillary zone electrophoresis. *Anal. Chim. Acta.* 377(1), 85-93 (1998).

[51] Ebrahimpour, B., Yamini, Y. and Esrafili, A. Emulsification liquid phase microextraction followed by on-line phase separation coupled to high performance liquid chromatography. *Anal. Chim. Acta.* 751, 79-85 (2012).

[52] De Orsi, D., Gagliardi, L., Porrà, R., Berri, S., Chimenti, P., Granese, A., Carpani, I., and Tonelli, D. A environmentally friendly reversed-phase liquid chromatography method for phthalates determination in nail cosmetics. *Anal. Chim. Acta.* 555(2), 238-241 (2006).

[53] Kamarei, F., Ebrahimzadeh, H. and Yamini, Y. Optimization of ultrasound-assisted emulsification microextraction with solidification of floating organic droplet followed by high performance liquid chromatography for the analysis of phthalate esters in cosmetic and environmental water samples. *Microchem. J.* 99(1), 26-33 (2011).

[54] Su, R., Zhao, X., Li, Z., Jia, Q., Liu, P., and Jia, J. Poly(methacrylic acid-co-ethylene glycol dimethacrylate) monolith microextraction coupled with high performance liquid chromatography for the determination of phthalate esters in cosmetics. *Anal. Chim. Acta.* 676(1-2), 103-108 (2010).

[55] Chingin, K., Chen, H., Gamez, G., Zhu, L., and Zenobi, R. Detection of diethyl phthalate in perfumes by extractive electrospray ionization mass spectrometry. *Anal. Chem.* 81(1), 123-129 (2009).

[56] Koniecki, D., Wang, R., Moody, R. P., and Zhu, J. Phthalates in cosmetic and personal care products: Concentrations and possible dermal exposure. *Environ. Res.* 111(3), 329-336 (2011).

[57] Feng, C.-H. and Jiang, S.-R. Micro-scale quantitation of ten phthalate esters in water samples and cosmetics using capillary liquid chromatography coupled to UV detection: effective strategies to reduce the production of organic waste. *Microchim. Acta.* 177(1-2), 167-175 (2012).

[58] ISO 12787:2011, "Cosmetics–Analytical methods–Validation criteria for analytical results using chromatographic techniques".

[59] Anon (1996) Seventh Directive 96/45/EC of the Commission. In: *Off. J. Eur. Commun.* L 213.

[60] Anon (2002) Commission Decision 2002/657/EC In: *Off. J. Eur. Commun.* L 221/8.

[61] Gao, W. and Legido-Quigley, C. Fast and sensitive high performance liquid chromatography analysis of cosmetic creams for hydroquinone, phenol and six preservatives. *J. Chromatogr. A.* 1218(28), 4307-4311 (2011).

[62] Gao, W., Gray, N., Heaton, J., Smith, N. W., Jia, Y., and Legido-Quigley, C. UV gradient combined with principal component analysis: Highly sensitive and specific high performance liquid chromatography analysis of cosmetic creams. *J. Chromatogr A.* 1228, 324-328 (2012).

[63] Yang, Y., Kapalavavi, B., Gujjar, L., Hadrous, S., Marple, R., and Gamsky, C. Industrial application of green chromatography – II. Separation and analysis of preservatives in skincare products using subcritical water chromatography. *Int. J. Cosm. Sci.* 34(5), 466-476 (2012).

[64] Wu, T., Wang, C., Wang, X., and Ma, Q. Simultaneous determination of 21 preservatives in cosmetics by ultra performance liquid chromatography. *Int. J. Cosm. Sci.* 30(5), 367-372 (2008).

[65] Mincea, M., Lupşa, I., Cinghiţă, D., Radovan, C., Talpos, I., and Ostafe, V. Determination of methylparaben from cosmetic products by ultra performance liquid chromatography. *J. Serb. Chem. Soc.* 74(6), 669-676 (2009).

[66] Mincea, M., Lupşa, I., Talpos, I., and Ostafe, V. UPLC Analysis of Common Parabens in Cosmetic Products. *Acta Chromatogr.* 2009, 4, 591-602.

[67] Cheng, Y.-C., Wang, C.-C., Chen, Y.-L., and Wu, S.-M. Large volume sample stacking with EOF and sweeping in CE for determination of common preservatives in cosmetic products by chemometric experimental design. *Electrophoresis*. 33(9-10), 1443-1448 (2012).

[68] Labat, L., Kummer, E., Dallet, P., and Dubost, J. P. Comparison of high-performance liquid chromatography and capillary zone electrophoresis for the determination of parabens in a cosmetic product. *J. Pharmaceut. Biomed. Anal.* 23(4), 763-769 (2000).

[69] He, S., Zhao, Y., Zhu, Z., Liu, H., Li, M., Shao, Y., and Zhuang, Q. Comparative study for the analysis of parabens by micellar electrokinetic capillary chromatography with and without large-volume sample stacking technique. *Talanta.* 69(1), 166-171 (2006).

[70] Wang, P., Ding, X., Li, Y., and Yang, Y. Simultaneous Determination of Eleven Preservatives in Cosmetics by Micellar Electrokinetic Chromatography. *J. AOAC Int.* 95(4), 1069-1073 (2012).

[71] García Jiménez, J. F., Carmen Valencia, M. and Fermin Capitán-Vallvey, L. Parabens determination with a hybrid FIA/HPLC system with ultra-short monolithic column. *J. Anal. Chem.* 65(2), 188-194 (2010).
[72] Ballesta Claver, J., Valencia, M. C. and Capitán-Vallvey, L. F. Analysis of parabens in cosmetics by low pressure liquid chromatography with monolithic column and chemiluminescent detection. *Talanta.* 79(2), 499-506 (2009).
[73] García-Jiménez, J. F., Valencia, M. C. and Capitán-Vallvey, L. F. *Anal. Chim. Acta* 594(2), 226-233 (2007).
[74] Tsai, T.-F., and Lee, M.-R. Determination of antioxidants and preservatives in cosmetics by SPME combined with GC–MS. *Chromatographia* 67(5-6), 425-431 (2008).
[75] Sanchez-Prado, L., Alvarez-Rivera, G., Lamas, J. P., Lores, M., Garcia-Jares, C., and Llompart, M. Analysis of multi-class preservatives in leave-on and rinse-off cosmetics by matrix solid-phase dispersion. *Anal. Bioanal. Chem.* 401, 3293-3304 (2011).
[76] Locatelli, M. Novel techniques as a tool to investigate on the triggering of biological targets. *Current Drug Targets.* 14(9), 937 (2013).
[77] Locatelli, M., Melucci, D., Carlucci, G., and Locatelli, C. Recent HPLC strategies to improve sensitivity and selectivity for the analysis of complex matrices. *Instrum. Sci. Technol.* 40(2-3), 112-137 (2012).
[78] Locatelli, M., Governatori, L., Carlucci, G., Genovese, S., Mollica, A., and Epifano, F. Recent application of analytical methods to phase I and phase II drugs development: a review. *Biomed. Chromatogr.* 26, 283-300 (2012).
[79] Haunschmidt, M., Buchberger, W., Klampfl, C. W., and Hertsens, R. Identification and semi quantitative analysis of parabens and UV filters in cosmetic products by direct-analysis-in-real-time mass spectrometry and gas chromatography with mass spectrometric detection. *Anal. Methods.* 3, 99-104 (2011).
[80] Koryta, J., Karel, S. and Koryta, J. *Ion selective electrodes*; Cambridge University Press: Cambridge, United Kingdom, 2009.
[81] Wang, L.-H. and Chin, C.-Y. Determination of sulphur in cosmetic products using ion-selective electrode. *J. Anal. Chem.* 62(7), 688-690 (2007).
[82] Wang, L.-H. Simultaneous quantitative determination of fluorine and sodium monofluorophosphate in oral hygiene products. *J. Cosmet. Sci.* 52(6), 399-405 (2001).

[83] Chiu, M.-H., Zen, J.-M., Kumar, A. S., Vasu, D., and Shih, Y. Selective cosmetic mercury analysis using a silver ink screen-printed electrode with potassium iodide solution. *Electroanalysis* 20(20), 2265-2270 (2008).

[84] Zen, J.-M., Kumar, A. S., Lee, S.-C., and Shih, Y. Microliter volume determination of cosmetic mercury with a partially cross-linked poly(4-vinylpyridine) modified screen-printed three-electrode portable assembly. *Electroanalysis* 19(22), 2369-2374 (2007).

[85] Shih, Y., Kumar, A. S., Zen, J.-M., and Hsu, J.-C. Detection of hidden mercury in cosmetic products by partially cross-linked poly(4-vinylpyridine) screen-printed electrode. *Bull. Chem. Soc. Jpn.* 78(12), 2130-2134 (2005).

[86] Wang, L.-H., Yai, H.-M. and Chen, M.-T. Determination of mercury in cosmetics on electropolymerized film-modified electrodes. *J. Food Drug Anal.* 4(2), 115-120 (1996).

[87] Wang, L.-H. Determination of lead, arsenic and mercury in cosmetic formulations at modified electrodes. *Int. J. Cosmet. Sci.* 16(1), 29-38 (1994).

[88] Shih, Y., Zen, J.-M., Kumar, A. S., Lee, Y.-C., and Huang, H.-R. Determination of the toxic lead level in cosmetic-hair dye formulations using a screen-printed silver electrode. *Bull. Chem. Soc. Jpn.* 77(2), 311-312 (2004).

[89] De Oliveira, M. F., Saczk, A. A., Neto, J. A. G., Roldan, P. S., and Stradiotto, N. R. Flow injection amperometric determination of persulfate in cosmetic products using a Prussian blue film-modified electrode. *Sensors* 3(9), 371-380 (2003).

[90] Chen, P.-Y., Luo, C.-H., Chen, M.-C., Tsai, F.-J., Chang, N.-F., and Shih, Y. Screen-printed carbon electrodes modified with cobalt phthalocyanine for selective sulphur detection in cosmetic products. *Int. J. Mol. Sci.* 12, 3810-3820 (2011).

[91] Ojani, R., Raoof, R. O. and Norouzi, B. An efficient sensor for determination of concentrated hydrogen peroxide based on nickel oxide modified carbon paste electrode. *Int. J. Electroanal. Sci.* 7(3), 1852-1863 (2012).

[92] Chiu, M.-H., Kumar, A. S., Sornambikai, S., Chen, P.-Y., Shih, Y., and Zen, J.-M. Cosmetic hydrogen peroxide detection using nano bismuth species deposited built-in-three-in-one screen-printed silver electrode. *Int. J. Electroanal. Sci.* 6(7), 2352-2365 (2011).

[93] Luo, P., Liu, J., Li, Y., Miao, Y., and Ye, B. Voltammetric determination of methylparaban in cosmetics using a multi-wall carbon nanotubes/nafion composite modified glassy carbon electrode. *Anal. Lett.* 45(16), 2445-2454 (2012).

[94] Cruz Vieira, I. and Fatibello-Filho, O. Biosensor based on paraffin/graphite modified with sweet potato tissue for the determination of hydroquinone in cosmetic cream in organic phase. *Talanta* 52(4), 681-689 (2000).

[95] Guo, S., Wu, X., Zhou, J., Wang, J., Yang, B., and Ye, B. MWNT/Nafion composite modified glassy carbon electrode as the voltammetric sensor for sensitive determination of 8-hydroxyquinoline in cosmetic. *J. Electroanal. Chem.* 655(1), 45-49 (2011).

[96] Yuanyuan, Z. and Song, L. The voltammetric sensor for sensitive determination of 8-hydroxyquinoline based on MWNTS/Nafion/GCE. *Optoelectron. Adv. Mat.* 5(9), 924-927 (2011).

[97] Gerlache, M., Senturk, Z., Viré, J. C., and Kauffmann, J. M. Potentiometric analysis of ionic surfactants by a new type of ion-selective electrode. *Anal. Chim. Acta* 349(1-3), 59-65 (1997).

[98] Kozan, J. V. B., Silva, R. P., Serrano, S. H. P., Lima, A. W. O., and Angnes, L. Amperometric detection of benzoyl peroxide in pharmaceutical preparations using carbon paste electrodes with peroxidises naturally immobilized on coconut fibers. *Biosens. Bioelectron.* 25(5), 1143-1148 (2010).

[99] Kumar, A. S., Ji, Y.-M., Sornambikai, S., Chen, P.-Y., and Shih, Y. Flow injection analysis of ellagic acid in cosmetic skin-whitening creams using a dendritic nanostructured copper-gold alloy plated screen-printed carbon electrode. *Int. J. Electroanal. Sci.* 6(11), 5344-5356 (2011).

[100] Chen, P.-Y., Ji, Y.-M., Luo, C.-H., Chen, Y.-S., and Shih, Y. Quantification of ellagic acid in cosmetic products by using a partially preanodized screen-printed carbon electrode coupled with flow injection analysis. *Anal. Methods-UK* 3(1), 205-209 (2011).

[101] Mielech-Lukasiewicz, K., Puzanowska-Tarasiewicz, H. and Niedzielko, A. Electrooxidation of some antifungal agents and their square wave voltammetric determination in cosmetics and pharmaceutics. *Anal. Lett.* 44(6), 955-967 (2011).

[102] Wang, Y., Cao, Y., Fang, C., and Gong, Q. Electrochemical sensor for parabens based on molecular imprinting polymers with dual-templates. *Anal. Chim. Acta* 673(2), 145-150 (2010).

[103] Baranowska, I. and Bijak, K. Differential pulse voltammetry in analysis of disinfectants 2-mercaptobenzothiazole, 4-chloro-3-methylphenol, triclosan, chloramine-T. *Cent. Eur. J. Chem.* 8(6), 1266-1272 (2010).

[104] Wang, L.-H. and Tsai, S.-J. Voltammetric behaviour of chlorhexidine at a film mercury electrodes and its determination in cosmetics and oral hygiene products. *Anal. Chim. Acta* 441(1), 107-116 (2001).

[105] Jùnior, J. B. G., Araujo, T. A., Trindade, M. A. G., and Ferreira, V. S. Electroanalytical determination of the sunscreen agent octocrylene in cosmetic products. *Int. J. Cosmet. Sci.* 34(1), 91-96 (2012).

[106] Wang, L.-H. Voltammetric behaviour of sunscreen agents at mercury film electrode. *Electroanalysis* 14(11), 773-781 (2002).

[107] Wang, L.-H. and Shu-Juan, H. Studies on the voltammetric behaviour of azo dyes and its determination in cosmetic products. *Russ. J. Electrochem.* 46(12), 1414-1418 (2010).

[108] Fogg, A. G., Barros, A. A. and Cabral, J. O. Differential-pulse adsorptive stripping voltammetry of food and cosmetic synthetic colouring matters and their determination and partial identification on tablet coatings and cosmetics. *Analyst* 111(7), 831-835 (1986).

[109] La Pera, L., Saitta, M., Di Bella, G., and Dugo, G. Simultaneous determination of Cd(II), Cu(II), Pb(II) and Zn(II) in citrus essential oils by derivative potentiometric stripping analysis. *J. Agr. Food Chem.* 51 (5), 1125-1129 (2003).

[110] Dugo, G., La Pera, L., Pollicino, D., and Saitta, M. Determination of selenium content in different types of seed oils by cathodic stripping potentiometry (CSP). *J. Agr. Food Chem.* 51(19), 5598-5601 (2003).

[111] Gil, E. D. S., Andrade, C. H., Barbosa, N. L., Braga, R. C., and Serrano, S. H. P. Cyclic voltammetry and computational chemistry studies on the evaluation of the redox behaviour of parabens and other analogues. *J. Braz. Chem. Soc.* 23(3), 565-572 (2012).

[112] Chang, M.-L. and Chang, C.-M. Voltammetric determination of sunscreen by convenient epoxy-carbon composite electrodes. *J. Food Drug Anal.* 9(4), 199-206 (2001).

[113] Ziyatdinova, G., Khuzina, A. and Budnikov, H. Determination of sterically hindered phenols and α-tocopherol by cyclic voltammetry. *Anal. Lett.* 45(12), 1670-1685 (2012).

[114] Ojani, R., Raoof, R. O. and Norouzi, B. An efficient sensor for determination of concentrated hydrogen peroxide based on nickel oxide carbon paste electrode. *Int. J. Electrochem. Sci.* 7(3), 1852-1863 (2012).

[115] Martins, I., Carreira, F. C., Canaes, L. S., De Souza Campos Junior, F. A., Da Silva Cruz, L. M., and Rath, S. Determination of parabens in shampoo using high performance liquid chromatography with amperometric detection on a boron-doped diamond electrode. *Talanta* 85(1), 1-7 (2011).

[116] Lai-Hao, W. and Shu-Juan, H. Design of a flow-through voltammetric sensor based on an antimony-modified silver electrode for determining Lithol Rubine B in cosmetics. *J. Autom. Method Manag.* art. No. 896978 (2011).

[117] Garcia, C. D., Ortiz, P. I. BHA and TBHQ quantification in cosmetic samples. *Electroanalysis* 12(13), 1074-1076 (2000).

[118] Wang, L.-H. and Wang, C.-C. Electrochemical oxidation of aminophylline at film platinum electrodes, and its determination in cosmetic and pharmaceutical products. *Microchim. Acta* 153(1-2), 95-100 (2006).

[119] Chang, M.-L. and Chang, C.-M. Simultaneous voltammetric determination of ascorbic acid and its derivatives in cosmetics using epoxy-carbon composite electrodes. *J. Food Drug Anal.* 13(3), 205-211 (2005).

In: New Developments in Analytical Chemistry ... ISBN: 978-1-63463-427-4
Editor: Breanna Granger

Chapter 2

BIOINSPIRED CHEMICALLY MODIFIED ELECTRODES FOR ELECTROANALYSIS

Ana Luisa Silva[1,*], Edilson Moura Pinto[2], Eduardo Ariel Ponzio[3], Eduardo Costa de Figueiredo[4] and Felipe Silva Semaan[1]

[1]Departamento de Química Analítica, Universidade Federal Fluminense, Outeiro de São João Batista, Rio de Janeiro, Brazil
[2]Departamento de Ciências Exatas e Sociais Aplicadas, Universidade Sagrado Coração, Bauru, São Paulo, Brazil
[3]Departamento de Físico-Química, Universidade Federal Fluminense, Outeiro de São João Batista, Rio de Janeiro, Brazil
[4]Departamento de Análises Clínicas e Toxicológicas, Universidade Federal de Alfenas, Alfenas, Minas gerais, Brazil

ABSTRACT

This chapter aims to present main aspects of biomimetic sensors, their obtainment and analytical applications, briefly revisiting fundamentals of voltammetry. These fields have experienced an

* Corresponding author: analuisa_als21@hotmail.com.

impressive growth in recent years; basic and applied researches on bioinspired chemically modified electrodes are opening, this way, a wide range of new specific applications. An electroanalytical biosensing device is a self-contained integrated unit capable of providing specific semi-quantitative, or even quantitative, analytical information by using a biological recognition element suitably retained in direct contact with an electrochemical transduction element. Such devices can be easily classified according to biological specificity-conferring mechanism, to physical-chemical signal transduction or to reaction category they monitor. The first well-established biomimetic sensor is the enzyme-based biosensor, in which different enzymes can be coupled to electrochemical platforms by many different ways in order to improve selectivity or/and sensitivity, such enzymes can be used as purified material or even in natural extracts, keeping the main advantages and reducing costs in many cases. Other interesting biological recognition mechanism exploited for electroanalytical purposes is antigen-antibody reactions, which leads to an immunosensor, a special kind of sensor in which the analytical device contains a biological material (antibody or antigen) immobilized onto its surface. Both described systems (enzyme-substrate and antigen-antibody) must be carefully assessed in order to achieve a responsive layer in which biochemical components can be entrapped without loss of activity or stability; in this sense many different immobilization approaches have been described. These bioelectro-analytical approaches are, in fact, limited to some natural or recombinant molecules, to the availability and stability of recognition elements and their capability of entrapment onto electrode surfaces, besides these costs of purified material are also a drawback. Aiming to overcome these limitations but keeping on mind the exploitation of biomolecular recognition principles, many studies have report the use of molecularly imprinted polymers (MIPs) as electrode modifier. The synthesis of such polymers may be reached from functional cross-linkable monomers in presence of a target/template molecule, being categorized according to the kind of occurring binding interactions (covalent or non-covalent) followed by a template removal step, vital for the recognition mechanism (since it allows the existence of complementary cavities). MIP-based sensors are in general, by means of sensitivity comparison, still inferior to biosensors. On the other hand, the great stability, low costs, and possibility of modeling different polymers based on many different templates; including molecules without natural biochemical recognition are still advantages that encourage scientists. Most of these applications are dependent on the related mass and charge transport properties. Thus, it becomes essential to understand the mechanisms of those transports in these materials, what can be carried out by electrochemical impedance spectroscopy (EIS), applicable for characterization, modelling and understanding interfacial processes involving the mass and charge transport on surfaces of modified electrodes. The analytical signal generated by the above-mentioned interacting systems can be translated by various electroanalytical methods, specially voltammetry and amperometry, being their main advantages the possibility of analyzing *in situ* smaller amounts of sample, in a shorter time.

INTRODUCTION

This chapter is neither a reference text, nor a complete description regarding a wide-field of expertise such as sensors in Analytical Chemistry; authors aim, in such piece, to present a brief and concise report, revisiting some main aspects of electroanalysis, passing through historical marks and experimental aspects, leading to some new perspectives and trends, specially on the exploitation of biochemical and/or biomimetical systems as chemical modifiers onto electrode´s surfaces.

A BRIEF OVERVIEW REGARDING HISTORY AND FUNDAMENTALS OF ELECTROANALYSIS

For centuries charge-transfer phenomena have raised questions; from Dalton, Lavoisier, and Rutherford, to Faraday, passing through Arrhenius, Debye, and Hückel, such phenomena have being intensively studied, not only by the perspective of basic science but also, and perhaps, mainly, by their possible applications. In such sense, it becomes impossible to not cite Jaroslav Heyrovsky, chemist who in the early nineteen's (1922-1923) developed the basis for polarography, starting a complete new brand of analytical technique which culminated in the Nobel Prize in 1959 (Bard & Faulkner, 1980; Wang, J., 2006; Ticianelli & Gonzalez, 1998; Schoulz, F., 2005; Bagotsky, V. S., 2006).

Heyrovsky studied the relationship among current flow (through a dropping mercury electrode - DME), applied potential (*versus* a reference electrode; at first electrochemical cells were developed by using two electrodes), and concentration of electroactive species (specially, at that moment, reducible cations) in solutions when a controlled external perturbation is imposed to such media, moreover, he observed that not only potential, but also current are directly correlated to the target specie and allows the obtainment of both qualitative and quantitative information by means of suitable interpretation of recorded profiles, so-called polarograms (Barek, J., 2002; Zuman, P., 2006).

This raises the main aspects of electroanalysis to be considered, being them currently applied: the triad: 1-) *chemical media* (supporting electrolytes, solvents, among others), 2-) *external perturbation* (electrodes and cells involved, applied technique), and 3-) *analyte* (structure, properties and behavior under certain circumstances). Such triad (Figure 1) is actually the platform for every modern electroanalytical study; in this chapter we will focus on the development of new chemically modified electrodes in order to enhance electroanalytical aspects, passing through their main advantages and limitations.

The use of liquid ultra pure mercury as electrode material, attractive at the beginning when reducible cationic metals and organic species were the focused, due to many reasons (such as toxicity, need for special purification steps, and so one) became gradually less interesting and limited for routine use. First electrochemical cells with two electrodes (a reference, and a working electrode) presented in many cases problems specially related to the current flow through the reference electrode what, in 1942, was corrected by building the first three-electrode potentiostat, starting a new segment with lots of progress in order to expand potentiostat´s capabilities. In the 1960´s and 1970´s, significant advances were achieved in voltammetry; advances in the theory, instrumentation, and the introduction of computer-controlled systems, leading to significant improvements on sensitivity, and expanding the repertoire of applications (Fisher & Kelley, 1966; Nlirnberg, H. W., 1979; Vyskočil, J. B. V., 2009).

Basically, the electrochemical cell, in which the voltammetric experiment is carried out, consists of the sample dissolved in a solvent, an ionic electrolyte, and electrodes: for three-electrode cells (the most common) we can name a working electrode, a reference electrode, and a counter electrode. In a typical three-electrode arrangement, the potential is applied to the working electrode with respect to a reference electrode, and an auxiliary electrode is used to complete the electrical circuit, avoiding current flow in the reference electrode. The working electrode, in general, made of noble metals (silver, gold, or platinum) or carbon (e.g., glassy carbon) can present different well-defined areas and shapes, parameters which can be changed and optimized for special applications (Sawyer, D. T., 1995; Brett & Brett, 1996; Scholz, 2005; Compton & Banks, 2007, Wang, J., 2006).

The auxiliary/counter electrodes are non-reactive high surface area electrodes, for which the platinum foils or wires are common choices.

Reference electrodes can be either Ag|AgCl electrode or Hg|$HgCl_2$ (saturated calomel electrode - SCE); sometimes, a simple silver or platinum wire is used in conjunction with an internal potential reference provided by ferrocene 0,5 – 1,0 mmol L^{-1} (used in organic or mixed organic-aqueous media in which an unknown liquid-junction potential raises as function of the application of internal aqueous reference solutions; such "probe" is added to the system in order to monitor potentials and shifts since its potential remains constant despite changes on chemical media), when a conventional reference electrode is not suitable.

Working electrodes, which make contact with the analyte, must apply the desired potential in a controlled way and facilitate the charge-transfer to/from such analyte, resulting in mass transport of new material to the electrode surface and generation of current.

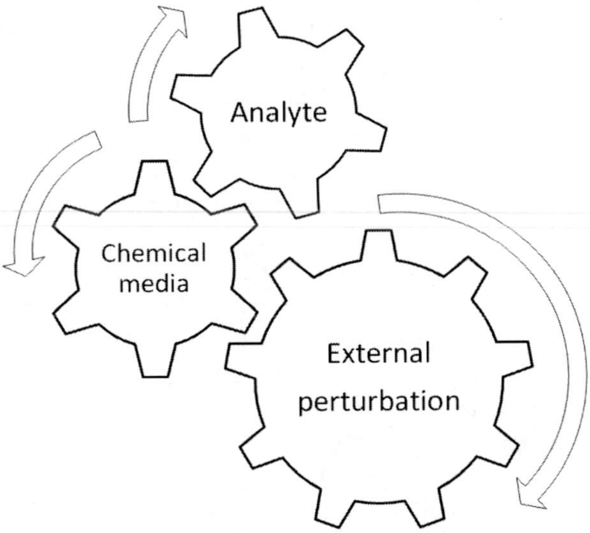

Figure 1. Schematic triad for electroanalytical experiments: (1) *chemical media* (supporting electrolytes, solvents, among others), (2) *external perturbation* (electrodes and cells, applied technique), and (3) *analyte* (structure, properties and behavior under certain circumstances).

IONIC EQUILIBRIUM AND ELECTROLYTES: BASIC IMPORTANT ASPECTS

In general, the most frequently used media for electrochemical experiments consist of ionic solutions, so-called electrolytes, solutions in which it is possible to assume not only dissolution but also ionization of solutes in a certain suitable solvent. This way, it becomes necessary to define solute and solvent, as well as their respective properties and influence on the media.

Solutes are, in general, components that, in minority, are solubilized in a defined solvent, in majority (except for the case of water, in which it is always the solvent). Their concentration as well as their interactions with such solvent will strongly influence several physical and chemical properties of the final solution. Such substances can be grouped into true (a) or potential electrolytes (b).

The first group (a) is related to those compounds that, in their pure solid state appear as ionic crystals which, in their molten state (pure liquid state), show to be ionic conductors. On the other hand, potential electrolytes (b), the second above mentioned group, show little conductivity even in their liquid pure form. This second group, thus, consist of separated neutral molecules with predominant covalent bounds among atoms, being this way strongly affected by the solvent. In other words, according to the basic properties of the solvent, such electrolytes can be either in a neutral or charged form (Bagotsky, V. S., 2006).

Solvents can be easily divided into two main groups according to their relative polarity: amphiprotic, which are capable of both accept and donate protons; and aprotic, which does not. They can be further categorized according to their protogenic or protophilic properties, and relative permissivity or polarity.

Polar solvents show molecules with permanent dipole moment which, under an electrical field, can be oriented against electrical field vector, generating this way a counter electric field; in some cases such behavior can even induce a dipole moment, hence, the net electrical field in such solvent becomes less than it is under vacuum. This decreasing of electrical field is proportional to an empirical constant so-called dielectric constant (ε), or permissivity of the solvent (Table 1) (Scholz, F., 2005; Bagotsky, V. S., 2006).

Table 1. Common solvents and their respective permissivity and autroprotolysis constants, under specified temperatures

Solvent	Permissivity of pure liquids[a]	Autoprotolysis equilibrium constants (pK_{ap})[b]
Acetic acid	6.15	14.46
Acetonitrile	37.5	26.50
Ethanol	24.3	19.10
Formic acid	57.9	6.20
Methanol	32.6	16.70
Water	80.2	14.00

[a] ε at 20°C.

[b] at 25°C, considering: $2HSol \Leftrightarrow H_2Sol^+ + Sol^-$

$$K_{ap} = \left[H_2Sol^+\right]\left[Sol^-\right]$$

The theory of solutions was strongly influenced by early studies of many scientists such as F. M. Raoult, J. H. van´t Hoff, M. Farady, S. Arrhenius, among others. Their conclusions guided all the subsequent theories, generating the modern theory of electrolytic dissociation. At the beginning colligative properties were studied, being defined as those properties that depend not on the chemical nature of solutes but on their concentration (Bockris & Reddy, 1970; Bagotsky, V. S., 2006). These findings point that, in contrast to nonelectrolytic solutions, in the case of electrolytic solutions, colligative properties depart appreciably from the values following from the equations above, even in highly diluted solutions that otherwise by all means can be regarded as ideal.

At early years of the nineteenth century, the first theories focused on the explanation of the two main features of electrolytic solutions: conduction, and the fact that chemical reactions could occur at electrodes during current flow. The first theory was proposed by T. Grotthus, in 1806, who introduced the concept that water molecules might be dipolar; according to his theory, under the effect of an electrical field water molecules will align in chainlike fashion, being the terminal chains, in contact to electrodes, responsible for evolution of hydrogen and oxygen by decomposition (Bockris & Reddy, 1970; Trassati, S. 1972; Bagotsky, V. S., 2006).

Besides this, another idea about the existence of independent charged particles in solutions emerged from M. Faraday´s studies (1834), who introduced some new terms such as ion, anion, cation, among others. He first pointed out that the moving ions at once secure the transport of electricity (charges) and the transport of the substance that reacts ate the electrode. Faraday assumed, however, that ions are formed from uncharged molecules only upon application of electric field, which gave rise to the term electrolyte (i.e. "*one unit or dissolved by electricity*"). The first ideas that ions might form spontaneously were formulated in the 1850s. In 1857, R. Clausius thought that ions could form spontaneously during collisions of solute molecules, but gave them a very short lifetime and assumed their fraction among the total number of molecules to be negligible.

S. Arrhenius developed a theory pretty much close to the modern concepts in 1883 in his doctoral thesis being published in 1887. He was the first to point out that conductivity and a departure of colligative properties from normal/expected values always occur together, concluding, this way, from his observations, that the two effects should have the same origin. Such observations could be summarized in three main points: 1-) in electrolyte solutions the molecules dissociate into ions spontaneously, turning the solution conductive; different electrolytes exhibit different degrees of dissociation, called α, which will influence actual values of molar conductivity Λ ($\alpha = \frac{\Lambda}{\Lambda^0}$), where Λ^0 is the limiting value of Λ at complete dissociation; 2-) due to the dissociation and the resulting increase in the total number of particles in solution, the parameters of the colligative properties assume higher values; 3-) for any given electrolyte dissolved, the degree of dissociation increases as the solution is diluted.

Few years later (1865 - 1887), D. Mendeleev proposed his theories regarding chemical solutions, according to these dissolution processes are chemical interactions between solvent and solute particles; upon salt dissolution in water, dissolved hydrates are generated. Later, in 1889, Mendeleev criticized Arrhenius´ achievements and this last, for himself, refuse to accept the existence of hydrates in solutions, both theories are, actually, complementary. Such existence of hydrates was, then, pointed out by A. Kablukov, in 1891 (Bockris & Reddy, 1970; Trassati, S. 1972; Bagotsky, V. S., 2006).

Special attention is often devoted to the water behavior for some many reasons. In aqueous media, which has a very high dielectric constant, the so-called strong electrolytes include all salts and strong acids and bases. Weak electrolytes are those with low ionization, such as weak organic acids and bases. Despite this, the distinction depends on the main properties of the solvent; it is possible to note solvents more acidic than water, while others are more alkaline. Besides these, permittivity of organic solvents tends to be smaller than that for water.

GENERAL BASIC ASPECTS OF VOLTAMMETRIC APPROACHES: EXTERNAL PERTURBATION

Electrochemical systems are those from which it is possible to note charge separation, often associated to charge-transfer. Such phenomena can occur spontaneously by means of conversion of energy obtained in a certain electrochemical reaction in electronic flow, characterizing this way the so-called galvanic cell. On the other hand, by offering to the system some external energy it is possible to stop or even invert such electronic flow, in such electrolytic system current is observed as function of applied potential, concentration of reactants, extension and time of reaction, among others. Such signal (current) can be divided into two main components to be discussed: 1-) faradayc (associated to charge-transfer between electroactive species and working electrodes), and 2-) capacitive (or non-faradayc, related to charge-discharge of electrical double-layer formed on the surroundings of electrodes) (Bard & Faulkner, 1980; Brett & Brett, 1996; Scholz, F., 2005; Wang, J., 2006, Compton & Banks, 2007).

Since electrochemical measurements are, in general, carried out in electrolytic media, many points must be considered at the moment of experimental planning and set up; the interaction among solvent and solute is just one of such parameters. The choice of solvent is primary based on electrolyte and analyte solubilities, in general, due to the high dielectric constant and other properties, water is often used as solvent, nevertheless other chemical media can also be applicable and interesting depending on some required properties.

Starting from now, considering many practical advantages, solutions must be considered prepared in pure water (except when clearly specified; in this otherwise case, mixtures between water and organic solvents must be carefully assessed, specially regarding to the possible effects on solubility, permissivity and autoprotolysis behavior, as described by Kiliç and Aslan, 2005). At this moment it becomes necessary to define the well-known term supporting electrolyte.

SUPPORTING ELECTROLYTES AND MASS-TRANSPORT MECHANISMS: CHEMICAL ENVIRONMENT X ANALYTE

Supporting electrolyte is a certain electrolyte, added to the system with many purposes; it can be used to control and adjust values of pH, viscosity, interfacial tension, ionic mobility and activities among others. The main function of such electrolyte is, clearly, to improve electrical conductivity of the system, which is reached by many mechanisms. Such supporting electrolyte must present suitable solubility, being inert on the chemical and electrochemical points of view (Agostinho et al., 2004).

By this point it becomes interesting to point out some aspects of solutions behavior. Up to now it was pretty simple to consider solutions as ideal, but, on the other hand, departure from such behavior can be observed, specially due to interactions between individual components (solvent and solute); ions in solution are subject to at least two types of forces, solvation, which is regarded to interactions with solvent particles, and electrostatic interactions with other ions. Interionic interactions tend to a negligible value when diluted solutions are exploited, specially due to a greater distance among charged particles ($F_{el} \downarrow = K \frac{q_1 q_2}{d^2 \uparrow}$); however, solvation occurs even in highly diluted systems since every ion is surrounded by many solvent particles. Such observations lead us to conclude that in electrochemical systems not always the approximation between concentration and activity is valid and such difference must be considered (Bockris & reddy, 1970; Scholz, F., 2005; Bagotsky, V. S., 2006).

An electrochemical cell consists of, at least, two electronic conductors (or electrodes) externally connected via a circuit and separated by an ionic conductor, called electrolyte. In voltammetry, the potential of the working electrode is controlled versus a reference electrode using a potentiostat, which allows the current flow between working and auxiliary (counter) electrode, in case of three-electrode cells, or even a current flow between reference and working electrodes, in case of two-electrodes cells (less common). Charge-transfer processes occur at electrode-solution interface, being related to both qualitative and quantitative aspects by means of parameters such as potential, current or even capacitance (Brett & Brett, 1996; Scholz, F., 2005; Wang, J., 2006).

The principle of ionic conduction is directly related to ion mobility, in liquid electrolytes it is consequence of a three-dimensional random movement of ions. Such movement allows a uniform concentration of ions throughout the solution in absence of external electric field by a mass-transport mechanism called diffusion. On the other hand, in case of application of external forces such as electric fields, ions acquire a nonrandom component of velocity directly proportional to such external influence (Brett & Brett, 1996; Ticianelli & Gonzalez, 1998; Scholz, F., 2005; Wang, J., 2006; Compton & Banks, 2007).

Obviously, charged particles will obey Faraday´s Laws and thus, influence the electrochemical properties of the solution, specially in polarized interfaces, such movement, called migration, will be well discussed later.

Let us, at this point, consider the diffusion and its influence on electrochemical systems. This process is caused by a gradient of concentration, a diffusional flux appears near the electrode surface where charge-transfer reactions take place, causing a depletion of ions, developing this way, such concentration gradient. The flux is defined by Fick´s First Law (1855):

$$J_D = -D\left(\frac{\partial c}{\partial x}\right)$$

While the concentration of ions at a determined distance from electrode surface is determined by Fick´s Second Law:

$$\frac{\partial c}{\partial t} = D\frac{\partial^2 c}{\partial x^2}$$

If the solution is not mechanically stirred, the diffusion is the only mode of mass-transport for uncharged electroactive particles and the dominant mode for ions in traces, the migration of which is negligible in the excess of supporting electrolyte.

Besides this, let us now consider another mass-transport mechanism. The velocity developed by a charged particle under unit applied force is called absolute mobility of the ion; the conventional, or electrochemical, mobility is the velocity of ions in a unit external electrical field, they are related themselves by:

$$\mu_{conv} = \mu_{abs} z_i e^0$$

In which $z_i e^0$ is the charge on the ion. The electrolyte contains at least two types of ions with opposite charges, to keep electroneutrallity obeyed, in liquids all ions are mobile and contribute to the global conductivity, providing that no ion-pairing occurs. The ionic conductivity is achieved by a nonrandom, direct movement of ions, resulting, this way, not only in mass-transport but also in flow of charge. Note that molecular compounds can easily distribute in different charged units depending on solvent (ε and pK_{ap}), final/effective pH (media) x pK_a (solute/analyte, Figure 2):

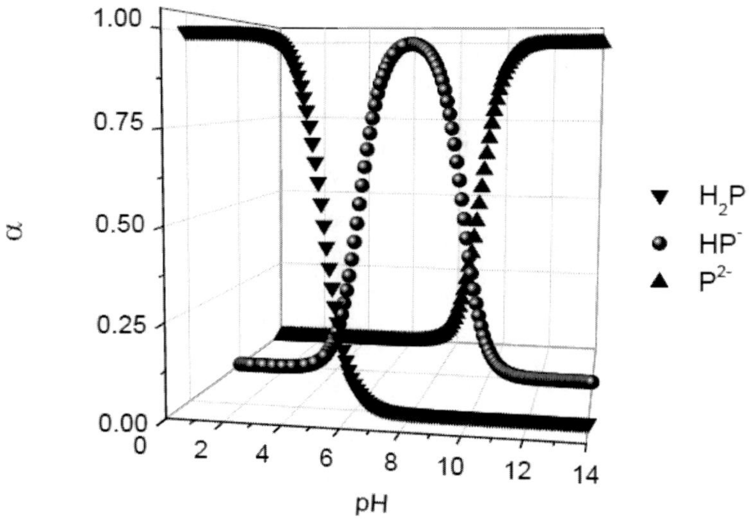

Figure 2. Example of species distribution as a reflex of chemical environment, consider a diprotic molecule H_2P (pK_{a1} = 5.2, and pK_{a2} = 9.8).

The migration is, thus, a flow of charge produced by an electric field between electrodes; cations moving to the cathode and, obviously, anions to the anode. In a fairly simple z:z-valent electrolyte the current density, or the flux of charge, is proportional to the concentration of ions, ($c = c^+ = c^-$, also $z = z^+ = z^-$), their conventional mobility (μ^+_{conv} and μ^-_{conv}) and the gradient of electrostatic potential, i.e., the electric field:

$$X = -\frac{\partial \psi}{\partial x}$$

$$\vec{j} = zFc(\mu^+_{conv} + \mu^-_{conv})X$$

At infinite dilution, according to those described by Ostwald, the equivalent conductivity of the electrolyte can be separated into the equivalent conductivities of individual ions:

$$\Lambda^0 = \lambda_+^0 + \lambda_-^0$$

Where $\lambda_+^0 = F(\mu_{conv}^+)^0$ and $\lambda_-^0 = F(\mu_{conv}^-)^0$, such approach is called law of the independent migration of ions. However, the conventional mobility of ions depends on the concentration of the electrolytes because of interionic interactions. The more ion concentrations are increased, the most intense their interactions become, which is negligible at lower concentrations. For this reason the equivalent conductivity of an electrolyte is lowered as:

$$\Lambda = \Lambda^0 - Ac^{1/2}$$

In the above-cited relationship A is an empirical constant, being this useful to concentrations up to 10^{-2} mol L^{-1}. Systems containing several electrolytes can be exploited, being the total current density made up of the sum of contributions of all ionic particles present:

$$\vec{j} = \sum_i zFc(\mu_{conv})_i X$$

Certainly, such approach is as simple as possible, since does not consider other aspects such as ion radio, permissivity of solvent, and others.

The supporting electrolyte, added in a concentration closed to hundred times higher than electroactive species, ensures electroneutrallity in the diffusion layer and suppresses the gradient of diffusion electric potential. Besides it conducts the current through the cell and prevents the concentration polarization of the auxiliary electrode, the surface area of which is usually much larger than the area of the working electrode surface (Brett & Brett, 1996; Scholz, F., 2005; Wang, J., 2006; Compton & Banks, 2007).

In terms of practical approach, some aspects must be also considered. In general, double-distilled water can be used in the majority of analysis, for trace analysis additional steps are required. Aprotic solvents, carefully purified and dried are used for the analysis of electrochemically generated radicals (Brett &

Brett, 1996; Ticianelli & Gonzalez, 1998; Scholz, F., 2005; Wang, J., 2006; Compton & Banks, 2007).

Oxygen is present in all solutions exposed to ambient atmosphere (close to 20% oxygen at 1 atm of pressure) in a concentration of circa 1 - 5 x 10^{-4} mol L^{-1}, and interferes in several reactions below 0 V (*versus* SCE), thus, depending on the experiment, such gas must be removed from the system, what can be done by many different procedures. The electrochemical reduction of dissolved oxygen usually occurs via two well-defined steps, the first corresponds to the formation of hydrogen peroxide, and the second, corresponds to the peroxide reduction, as follows:

$$O_2 + 2H^+ + 2e^- \longrightarrow H_2O_2$$

$$H_2O_2 + 2H^+ + 2e^- \longrightarrow 2H_2O$$

$$O_2 + 2H_2O + 2e^- \longrightarrow 2OH^- + H_2O_2$$

$$H_2O_2 + 2e^- \longrightarrow 2OH^-$$

The half-wave potentials for all these steps, *versus* SCE, are complied between -0.1 and -0.9 V, respectively. The exact stoichiometry of both steps is clearly dependent on pH. The resulting background in presence of oxygen interferes with measurements of many reducible analytes; moreover, products of oxygen reduction can also affect other electrochemical processes under study.

Oxygen removal can be carried out by purging them with high-purity nitrogen (aqueous solutions) or even argon (organic media), for about 5 - 10 minutes prior the analysis, and by keeping a blanket of the same atmosphere over the solution to avoid oxygen reentering. Certainly such time must be suited to sample volume and oxygen contents. Other procedures include the use of electrochemical or chemical scrubbers, nitrogen-activated nebulizers, and chemical reduction (by addition of hydrazine, ascorbic acid, sodium sulfite, among others) (Trassati, S., 1972; Brett & Brett, 1996; Scholz, F., 2005; Wang, J., 2006; Compton & Banks, 2007).

Besides the interference of oxygen, another important point to consider is the possibility of water electrolysis, altering pH and generating bubbles on electrodes surfaces. In pure water the potential window is limited by reductive and oxidative decompositions (Trassati, S., 1972):

$$2H_2O + 2e^- \rightleftarrows H_2 \uparrow + 2OH^- \qquad (a)$$

$$2H_2O \rightleftarrows O_2 \uparrow + 4H^+ + 4e^- \qquad (b)$$

Since those reactions are reversible and inter correlated to pH values and chemical environment, corresponding Nernstinian relationships for simpler systems can be written and presented as (Figure 3).

$$E_{H^+/H_2} = -0.0592\,pH\,(vs\ SHE)$$

$$E_{O_2/H_2O} = 1.23 - 0.0592\,pH\,(vs\ SHE)$$

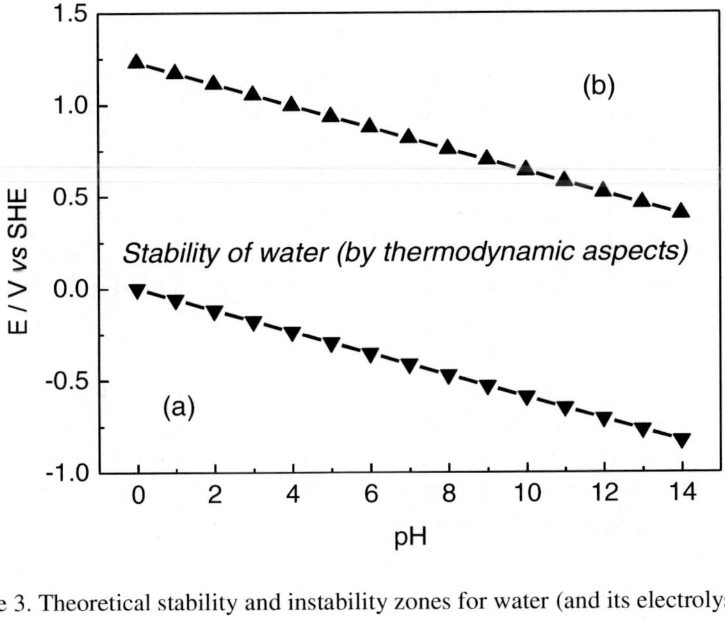

Figure 3. Theoretical stability and instability zones for water (and its electrolysis) under specified E (V) and pH conditions.

The potential range of electrode polarization is, actually, dependent not only on the solvent and electrolyte, but also the electrode composition; in the above-mentioned case consider a Pt working electrode.

Double Layers and their Influence on Electroanalysis: Chemical Environment *x* External Perturbation

By sinking any electrode into an electrolytic solution an interfacial region, with particular properties is formed, this so-called double layer can strongly interfere on surface phenomena and thus, it influences many electrochemical events under study. This special region at the surroundings of the electrode can be qualitative described by many different and often complementary models, since double layer structure, dimensions and capacity depend on several aspects from which we can highlight electrode material, nature of solvent and supporting electrolyte (as well as their concentrations), adsorption phenomena, among others, in other words, there is no general model that satisfies every situation found.

Historically, the first clue for the existence of such double layer emerged from Helmholtz´s studies, in 1879; this first theory assumed the presence of a compact layer of ions in direct contact with the charged electrode surface.

This first proposal was followed by the concepts of Gouy and Chapman (1913), who adapted the previously described model to another in which ions would be accumulated in a diffuse double layer, being such distribution a reflex of the Boltzmann distribution from the electrode surface to the bulk of the solution. Both of these concepts were then complied in one unique proposal by Stern (1924), in this case the polarized solid-liquid interface would include both the rigid Helmholtz plane and the diffuse layer of Gouy-Chapman.

Later, in 1947, Graham pointed that, since specific ions can adsorb onto electrode surfaces, such capacity could be directly affected. Up to this point no solvent role was not even cited but, further studies mainly by Parsons (1954), and Bockris, Devanathan and Muller (1963) had taken such approach into account (Bockris & Reddy, 1970; Bard & Faulkner, 1980; Brett & Brett, 1996; Ticianelli & Gonzalez, 1998; Scholz, F., 2005; Wang, J., 2006; Bagotsky, V. S., 2006; Compton & Banks, 2007).

This evolution pointed that in dipolar solvents (e.g. water), dipoles must interact with charged electrodes, being important to relate such interaction with solvent properties such as dielectric constant and so one.

Despite all particularities, let us now point some basic concepts regarding such double layer as well as its influence over electrochemical measurements. Basically this region is composed by two planes, the inner Helmholtz plane (IHP) is that passes through the centers of adsorbed ions or is just located behind the layer of adsorbed water, being also called compact layer; the second one, the outer Helmholtz plane (OHP), passes through the centers of hydrated/solvated ions in contact of electrode surface, both these planes are related to different electric potentials and generate certain capacity. Beyond the OHP the diffuse layer is observed. The concentration of ions in the diffuse layer varies as function not only of such electric potential but also regarding the distance from electrode surface (Bockris & Reddy, 1970; Bard & Faulkner, 1980; Brett & Brett, 1996; Ticianelli & Gonzalez, 1998; Scholz, F., 2005; Wang, J., 2006).

The capacity generated by the double layer is responsible by the rising of non-faradayc currents, which must be considered and excluded from some data for analytical purposes; many strategies have been described to isolate, minimize or even suppress such capacity current. The thickness of the double layer also depends on the concentration of electrolyte and electric potential.

Working Electrode Materials

The electrochemical system´s performance is strongly influenced by the working electrode material. The working electrode should provide high signal-to-noise characteristics, as well as a reproducible response.

The electrochemical behavior can vary strongly according to the material used for such purpose. There are many evidences that rate of electron transfer across an electrode-solution interface is dependent on the physical and chemical properties of electrode materials (Sawyer et al., 1995). The electrocatalysis term has been raised from some studies (1970 decade), providing wide evidence for such relationship electrode material *vs* electrochemical behavior (Bockris & Reddy, 1970; Trassati, S., 1972).

Reactions on the surface of an electrode are analogous to any other heterogeneous catalytic reaction which takes place on catalytic surfaces, but with a difference. Heterogeneous reaction does not involve a net charge transfer across the interface, therefore potential is independent. Reactions on electrode surfaces involve charge transfers across the interface electrode-solution, where the activation energy of the electrode reaction can be controlled by varying the potential (Bard & Faulkner, 1980; Brett & Brett,

1996; Ticianelli & Gonzalez, 1998; Scholz, F., 2005; Wang, J., 2006; Compton & Banks, 2007).

Many different materials have been presented for application as working electrodes in electrochemistry. Among the most popular materials used, those with higher importance are composed by mercury, or carbon, or noble metals (particularly gold and platinum). The working electrode shall comply some basic requirements: being stable on accessible potential window for electrochemical behavior of the target compound, suitable electrical conductivity on electrolyte, reproducibility and robustness, strength, costs and availability. Another important feature to be elucidated is the working electrode geometry (Bard & Faulkner, 1980; Brett & Brett, 1996; Ticianelli & Gonzalez, 1998; Scholz, F., 2005; Wang, J., 2006).

Most metals used as working electrode are unstable anodes and thus unsuitable for use in oxidative processes. The activity and selectivity of metals are, basically, determined by their chemical nature, with hardly possible changes.

Solid Electrodes: Main Platforms for Modifications

While for mercury electrodes surface roughness has a value of one (geometric/electroactive surface ratio equals one), the catalytic working area for other electrodes presents other values; consequently, for the use of such electrodes it is required different procedures to obtain accurate and reproducible results. The surface of a solid electrode is determined by the chemical nature of its material, and this feature derived from the polycrystalline and/or amorphous structure. Even polycrystalline materials have gaps between the grain boundaries, resulting in a surface area different from real electroactive surface (Brett & Brett, 1996; Scholz, F., 2005; Wang, J., 2006).

Working electrodes for use in electrochemistry are normally solid, with rare exceptions; mercury electrodes are one of those in which the electrode material is liquid at room temperature.

In general, electrode materials are conductors or semiconductors constituted by the above-cited substrates (metals, carbon and derivates) or even by some polymers; rarely metal oxides (despite their special use in some cases). Typically working electrodes are constructed in cylindrical form for manufacturing, with connections made by conductive silver glue and a metal wire/rod as copper, avoiding an exposure of such conductive metal with the electrolyte (Figure 4). After preparation, the obtained electrode should be polished before use to avoid physical imperfections; such polishment can be done by using suitable polishment materials such as alumina/diamond slurry, special sandpapers and so one, available in many different sizes (Brett & Brett, 1996; Scholz, F., 2005; Wang, J., 2006).

Prior to analytical use, steps of pretreatment will depend on the material and experimental conditions. Pretreatment for metal electrodes are usually with polishing and potential cycling. While carbon based electrodes need chemical/electrochemical treatments, or even thermal activation (Wang et al., 1990; Wang, J., 2006).

METAL ELECTRODES

In the academic literature has been written very about solid metal electrodes. Often noble metals are used because exhibit characteristics of being inert variation of applied potential but any metal electrode can be applied if it not is exceeded the limits of the electrochemical window on electrolyte solution.

In general platinum and gold are the most widely used metallic electrodes. Present favorable electron transfer kinetics and a large anodic potential range. However, for use on cathodic potential (-0.2 V until -0.5 V) is limited by low hydrogen overvoltage and formation of oxide on the surface. The film to oxide can be eliminated with to cleaning–reactivation cycles (Johnson & La Corse, 1960). The problem is less severe in nonaqueous electrolyte, but in nonaqueous media platinum can present catalytic characteristics (Scholz, F., 2005; Wang, J., 2006).

As a consequence of the narrow potential window and higher costs, noble metal solid electrodes are, often, the second choice as working electrodes, being used when failures on the methodology with carbon electrode occurs (Navratilova & Kula, 2000; Hocevar et al., 2005).

Gold electrodes are more inert to formation of oxide films that platinum electrode, on the other hand it can easily suffer corrosion in the presence of CN^- or Cl^-; considering such aspects and costs, gold electrodes are still a very useful tool for modifications by deposition, specially regarding thiol-containing components. Another metal have been used as electrode for the formation to oxihydroxide *in situ* for detection to amino acids or carbohydrates. Among this metal stand out copper, silver and nickel that, unlike gold and platinum, can present the stable response desirable (Scholz, F., 2005; Wang, J., 2006).

Another special circumstance regarding the use of gold electrodes is the fact, like silver, gold surfaces can spontaneously generate ordered self-assembled layers (SAM, self-assembled monolayers) when in direct contact to alkanethiols; such chemisorption is due sulfhydryl groups reaction generating gold-sulfur bonds and represents a great tool for the immobilization of enzymes, molecularly imprinted polymers, among others. In certain cases antibodies or their fragments after enzymatic digestion antibodies can expose disulphide bridges, generating chemisorption and thus allowing stable immobilization.

Carbon Based Electrodes

Carbon electrodes are widely used currently in electroanalysis, being the first choice primarily because of their broad potential window, low cost, low background current, chemical inertness, suitable for different types of sensing, despite the fact of its electron transfers are not as fast as those for metals (Wang, J., 2006). Electron transfer is highly affected for structure reactivity on surface carbon and numerous studies are devoted for their understanding (McCreery, R. L., 1991; Chen & McCreery, 1996).

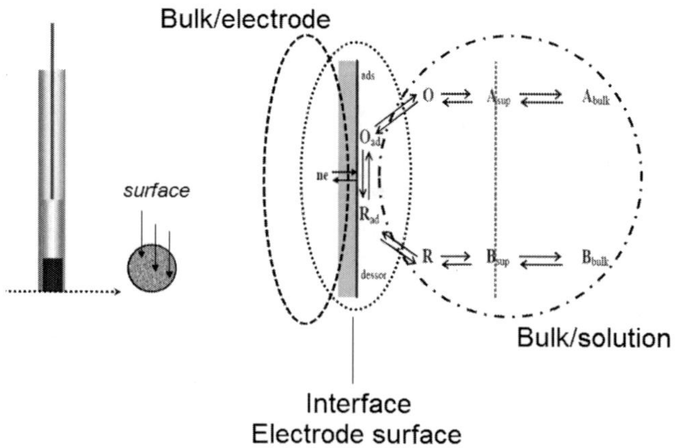

Figure 4. Main regions for electroanalytical purposes, chemical modifications can be carried out onto electrode surface as well as in bulk electrode, special focus will be devoted to the first option.

Carbon electrode materials are structured based on graphene (a six-member aromatic ring with sp^2 bonding). Appropriate procedures for surface cleaning and pretreatment are factors that provide increasing electrochemical reactivity at carbon electrodes and, thus, electron transfer rates. Sometimes the electrode should be suitable pre-oxidized by sulfuric solutions by applying currents circa 25 μA for few minutes, enhancing such way the hydrophilicity of the surface and, thus, the adsorptive properties, which could be extremely attractive on developing chemically modified electrodes.

Other variable present in carbon electrodes are pores on surface catalytic include increase additional frequency dispersion. These characteristics such as distribution heterogeneities, roughness, distribution of pores and increasing ability to anchoring, deposition or contamination of materials on the surface, should be considered depending on the process that will be conducted. In general such cavities and surface effects can be minimized by impregnating pores with paraffin, waxes, and so one, what reduces background currents (Scholz, F., 2005; Wang, J., 2006).

Among all carbon-based materials the most common and used is the glassy/vitreous carbon electrode. Commonly used as well, specially for certain applications (chemically modified electrodes, fast scan analysis) are carbon composite electrodes (also called carbon paste electrodes, e.g. graphite-mineral oil, graphite-paraffin, graphite-epoxy, graphite-polyurethane, graphite-paraffin), carbon fibers and doped diamond electrodes.

Glassy Carbon Electrodes (GCE)

Glassy carbon electrodes are very popular by present well-known suitable mechanical and electrical properties. Moreover their wide potential window, chemical inertness (solvent resistance), and reproducible performance turns them interesting. This material is prepared by a controlled carbonization process of phenol-formaldehyde resins (slowly, over the 300–1200°C in an inert atmosphere) (Adams, R. N., 1969). The obtained structure presents high density and small pore size.

Usually is employed a pretreatment on surface, achieved by polishing until mirror-like aspect (with different and gradually smaller particles sizes in sandpapers or alumina slurries), to improve elctrochemical performance (Engstrom, R. C., 1982; Engstrom & Strasser, 1984). Other pretreatments as activation with electrochemical steps, heat, or laser treatments have been also used to enhance performance (Wang et al., 1990; McCreery, R. L., 1991; Chen & McCreery, 1996). The improvement can be attributed to the exposure of graphene sheets, and removal contaminants. Reviews with more information regarding properties of glassy carbon electrodes are described in (Van der Linden & Dieken, 1980; McCreery, R. L., 1991; Chen & McCreery, 1996).

Other similar vitreous carbon material is reticulated vitreous carbon (RVC), a material sponge-like material (with increased surface catalytic properties) which has demonstrated quite a potential in applications such as flow analysis and spectroelectrochemistry (Arredondo Valdez et al., 2012).

CARBON COMPOSITE ELECTRODES (CARBON PASTES, CPE)

Carbon paste electrodes are hybrid materials which use graphite powder mixed with aggregating no conducting, have shown great interest in electroanalysis (Kalcher et al., 1995; Svancara et al., 2009). Such composites are viewed with great interest specially due to their edge effects and mechanical properties. Such materials are prepared by mechanical dispersion of suitable amounts of graphite powder and a binder liquid, and their choice should suit the low volatility and purity. After are prepare carbon paste, the composite has characteristics different to components originally separated, being considered as a new material (Adams, R. N., 1958; Tallman & Petersen, 1990). Carbon based materials represents, thus, a convenient matrix and has shown great interest in electrode modifications and construction of biosensors (Taleat et al., 2014).

Simple examples of insulating phases can be highlighted in the preparation of carbon paste electrodes: mineral oil (Nujol®) (Svancara et al., 2010), paraffin (Perantoni et al., 2011; Santos et al., 2013), silicone rubber (Oliveira et al., 2008), and different resins such as epoxy (Pauliukaite et al., 2009) and polyurethane (Cervini et al., 2002; Semaan et al., 2008). The composition of the hybrid strongly affects the electrode reactivity. Preliminary studies must be conducted in order to monitor the catalytic surface electroactive *versus* percentage of non-conductive support (due to the dependence of the electron transfer rates, as well as the background current contributions, on the composition of hybrid) (Rice et al., 1983).

Carbon Fiber Electrodes (CFE)

Studies using ultramicroelectrodes have been shown of great interest for electroanalysis, but initially the material for use on construction was known as fiber carbon. Such materials are prepared by carbonization on controlled high temperature of the polymer textiles or other means of obtainment is via catalytic chemical vapor deposition. There are three types of carbon fiber with depending on the manufacturing process. Exist three types of carbon fiber with depending on the manufacturing process and are classified as: low-, medium-, and high-modulus (Wang, J., 2006). Among the types presented the most suitable for electrochemical studies is high-modulo presenting the exposure of sheets of graphene, and low porosity (Edmonds, T. E., 1985; Feng et al., 1987; Csöregi et al., 1993).

Considering the formation of electric double layers onto electrode surfaces and their influence on faradayc and non-faradayc (capacitive) currents, clearly such ultramicroelectrodes allow fast scan voltammetric measurements, specially required under certain circumstances, for example, for the analysis of analytes with much high peak potential onto carbon surfaces (e.g. caffeine).

Doped Diamond Electrodes (DDE)

In nature the diamond structure isn´t capable of electrons transfer, it is an electrical insulator material. However, by modification techniques, boron-doped diamond films can show increased electron transfer characteristics, turning it into a material compared to semiconductors, presenting, without any pretreatment, a wide potential window (approaching +3 V), good electrochemical reactivity, mechanical hardness, and low adsorption on surface (in certain cases), such electrodes are highly useful for electrochemical measurements in higher potentials and under extreme conditions (media) (Brett & Brett, 1996; Terashima et al., 2002; Scholz, F., 2005; Wang, J., 2006).

CHEMICALLY MODIFIED ELECTRODES (CME): SEEKING NATURE´S RECOGNITION CAPABILITIES

Up to now we have focused some efforts on a brief but suitable "*jogging*" around electroanalytical principles but, in fact, the main goal of the present chapter is to revisit and present some possible ways to exploit chemical interfacial modifications in order to reach better analytical performance. In general, such improvements can be granted by following nature's mechanisms and processes of recognition.

Since the early 60´s electrochemical detection has been studied, being some of the first chemically modified electrode´s approach described by Clarks and coworkers, at that time an remarkable improvement on electrode´s analytical aspects were reached by modifying surfaces by enzymes (Chaubey & Malhotra, 2002; Turner, A. P. F., 2013). Later, Murray (1975) introduced the expression "*surface modification*" while describing chemical surface modifications onto SnO_2 electrodes by using silane, amines, and pyridil compounds. At first, and up to now, some were, and still are, the reasons for these modifications: stability and life-time improvement, stability to certain more aggressive applications, enhancement on sensitivity and selectivity, pre-concentration of analyte and exclusion of interference, among others (Souza, M. F. B., 1996; Pereira et al. 2002).

In general, it is practical to divide the main surface modification procedures or categories into four groups: 1-) physical sorption (or sometimes called adsorption, it is the adherence of biological material by dissolving the modifying agent in a suitable solvent, the surface of an electrode), 2-) covalent binding, 3-) thin film depositions (polymeric matrixes, metallic films, among others), and 4-) composites preparations (Souza, M. F. B., 1996; Kutner et al., 1998; Pereira et al., 2002). Besides such classification, biosensors can also be categorized by means of their transduction principles such as conductometric, potentiometric, voltammetric, amperometric, mediated (or not), among various (Souza, M. F. B., 1996; Pereira et al. 2002; Chaubey & Malhotra, 2002).

In brief, the definition of an electrochemical biosensor by the International Union of Pure and Applied Chemistry - IUPAC cites that "*electrochemical biosensor is an integrated self-contained instrument, capable of providing specific quantitative or semi-quantitative analytical information by using a biological recognition element (receptor) in direct spatial contact with a transduction element*" (Thevenot et al., 2001), as basically depicted in Figure 5. Such analytical device gathers remarkable relationship among sensitivity and specificity found in biological systems to physicochemical transducers to provide complex analytical measurements in a simple way, allowing virtually every conceivable analytical approach (Turner, A. F. P., 2013).

The transducer is, thus, responsible for the detection of changes in the interfacial physical and/or chemical behavior as a function of analyte´s presence, being such analytical signal detected by direct (using a biological marker) or indirect methods (no marker is required) (Liu et al., 2012; Liu et al., 2013).

For the exploitation of biosensors it becomes necessary to meet certain requirements, including the choice of a biological (or sometimes biomimetic, bioinspired) material (pure or even in crude extracts), with active sites or, at least, possibility to incorporate biological material (extracts, cells, tissues) with active sites capable of interacting or reacting with the chosen analyte (modification or immobilization), and chemical media (environment) that meets the conditions of stability needed for both enzyme-substrate-measurement (Figure 5)(Gonzalez et al., 2004).

Considering the above-mentioned overview it is possible to conclude that wide is the applicability of such devices depending on the recognition component: enzymatic biosensors, immunosensors, MIP-based sensors, among others (such as DNA biosensors, which will not be mentioned here, despite their importance).

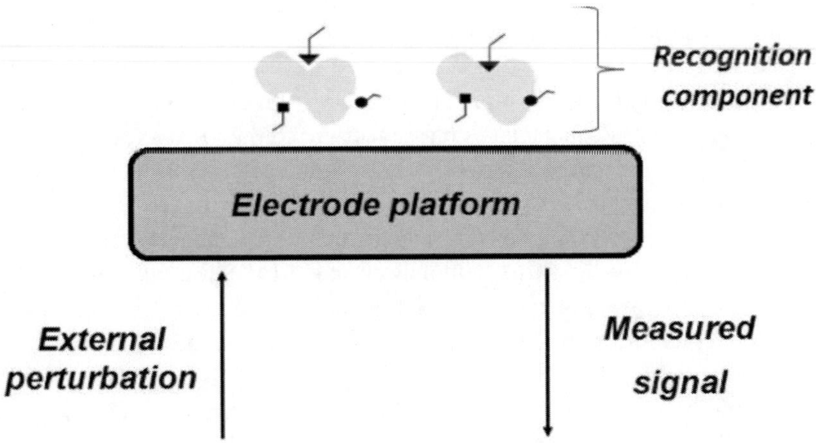

Figure 5. Scheme for a simple biosensor: an electrodic platform, a recognition component layer, external perturbation and signal measurement.

Enzymatic Biosensors

A simplest enzyme electrode consists of a thin layer of immobilized enzyme (pure or in a crude extract, cells or even tissues) held onto a suitable platform surface (acting as a transducer), such electrode is thus used in association to a counter electrode (in general, Pt wires or foils) and a suitable reference electrode, providing this way a three-component circuit for measuring either by potentiometry or amperometry, analytical responses in this case are typically dependent on the amount of active enzyme immobilized (Souza, M. F. B., 1996; Pereira et al. 2002; Chaubey & Malhotra, 2002).

While carrying out the measurements, the enzyme electrode is immersed into a suitable chemical media (for example, buffers such as phosphate, acetate, organic-water mixtures, among others) containing the analyte, being steady state potential (E) or current (*I*) read. A logarithmic relationship is often observed for the potentiometric, being the response linear for amperometric/voltammetric measurements. For a kinetically controlled biochemical reaction catalyzed by the immobilized enzyme, the steady state current (*I*) is proportional to a initial rate of enzymatic process (*k*). In this case a plot of *I versus* substrate concentration leads to a typical Michaelis-Menten type response, what can be defined by means of assessing a linear 1/I *versus* 1/S plot (Turner, A. F. P., 2013).

Different procedures for enzyme immobilization has been described and widely discussed. For didactic purposes such step on biosensor construction can be summarized in: entrapment (electropolymerization of conducting polymers, amphiphilic matrixes, entrapment in sol-gel matrixes, or even in composites such as carbon pastes, clays, among others), adsorption (by physical processes, electrostatic interactions, ion-exchangers, layer-by-layer processes, adsorption onto metal and metal oxide nanoparticles, among others)(Ansari & Husain, 2012; Cipolati et al., 2014; Shi et al., 2014), cross-linking (in many different matrixes such as chitin, chitosan, melamine, using agents as epichloridrine, glyoxal, glutaraldehyde)(Krajewska, B., 2004; Pauliukaite et al., 2009: Sowmya & Meenakshi, 2014), covalent bonding (chemisorption by activation of carboxylic and/or amino groups)(Yang, H., 2012), bioaffinity interaction (biotin-streptavidin, lectin-carbohidrate)(Sassolas et al., 2012).

Each immobilization procedure presents advantages and drawbacks, being the one´s choice dependent on the enzyme, the transducer and the technique to be applied. Besides these, such choice also must be carried out by considering other requirements such as sensitivity or stability, reproducibility, costs and operational difficulties (Sassolas et al., 2012; Cipolati et al., 2014).

Table 2 presents some interesting examples of applications of the above-mentioned enzyme-based biosensors.

IMMUNOSENSORS

Immunosensors are a subtype among biosensors in which the recognition system based on immunological reactions (antigen-antibody or Ag-Ab) is suitably immobilized onto platform surfaces connected to a transducer (which produces a quantitative signal capable of measuring the analyte concentration, Figure 6) (Healy et al., 2007).

The current challenge for industries, pharmaceutical companies, clinical interests, such as for the diagnosis of diseases is the search for more sensitive, rapid and selective methods, thus, the development of new technologies capable of replacing sophisticated devices by miniaturized portable units, easier to handle and cost-affordable has become a new and promising research field.

Immune interaction has been analytically exploited in various techniques such as enzyme-linked immunosorbent assay - ELISA, the major drawback of these techniques is the difficulty of adapting them to field and/or *in situ* analysis. The use of immunosensors based on ELISA has been one of the main alternatives to detect specific antibodies or antigens for the diagnosis of various diseases.

Table 2. Some current and interesting examples of applications for enzymatic biosensors

Analyte	Technique, detection	Reference
Cellobiose	Amperometric	Cruys-Bagger et al., 2012
Dichlorvos, aldicarb e paration	Amperometric	Ito et al., 2014
Glicose	Cyclic voltammetry	Ekabutr et al., 2013
	Amperometric	Li et al., 2012
	Amperometric	Demirkiran et al., 2012
L-lysine	Amperometric	Ciriello et al., 2015

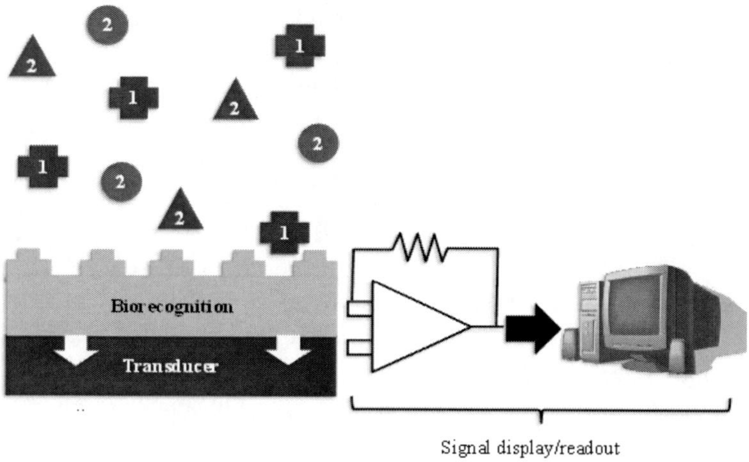

Figure 6. Scheme of the basic components for an immunosensor and its possible detection techniques, consider (1) the analyte.

The choice of the sensor layer is a parameter of paramount importance, since it must take into account the interaction of high specificity and selectivity between antigen and antibody, making them chemically-active parts of the transduction, turning this way, such interaction a source of measurable analytical signal.

The specificity of the interaction between Ag and Ab (as depicted, Figure 7), is directly related to the fragment antigen binding - F_{ab}, being the interaction located at the epitope portion, and forming the Ag-Ab complexes. This interaction follows an affinity constant related to the concentration of the species Ag and Ab, and expressed by:

$$Ag + Ab \Leftrightarrow Ag : Ab$$

$$K_{aff} = \frac{Ag : Ab}{Ag \quad Ab} \geq 10^8$$

As described in the previous section, some requirements must be met for the development of an immunosensor, and it was mentioned that there should be an interaction between antibody and antigen to be placed onto electrode platform, for that some immobilization methods are used, including adsorption, covalent linkage, covalent cross-link, occlusion, among others.

The main advantage of adsorption is grip the material in a very wide variety of electrodes, being, thus, one of the most used techniques due to its simplicity, efficiency; it also causes less damage to biomolecules (Sharma et al., 2003).

Surface modifications by covalent bonding is, in general, very stable when compared to others, however, although it generates only an immobilized monolayer being, thus, considered the ideal method to modify many electrodes, facilitating a possible commercialization (Gerard et al., 2002).

In the case of covalent cross-links, the biological component is reacted with a bifunctional agent to form intermolecular covalent bonds (net-like). For the choice of such bifunctional agents, it should be assessed the structure of the biological component, since those interactions depend on functional groups present in the target biomolecule (Malhorta et al., 2006). Different cross-linking reagents are currently available and their use is well-known (Pauliukaite et al., 2009).

The occlusion method is simple and consist of prepare solutions containing both electrode surface modifier and biological recognition component, allowing them to intercalate and remain entrapped. It is a process involving both physical interaction or covalent bonding (Gerard et al., 2002; Malhotra et al., 2006).

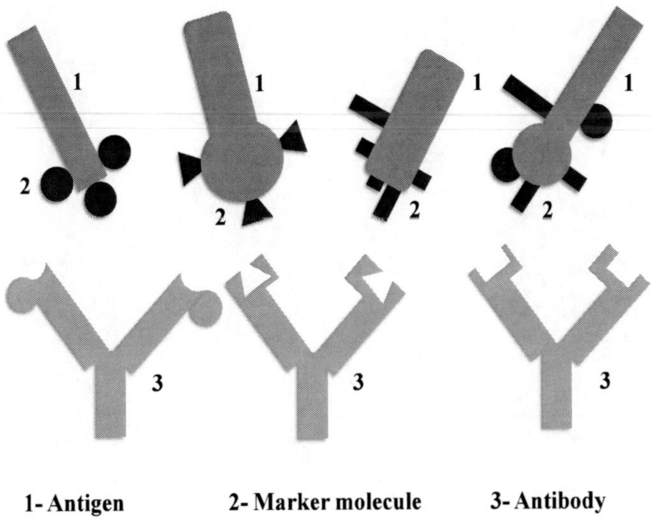

Figure 7. Particular combination between an antibody and its respective complementary antigen.

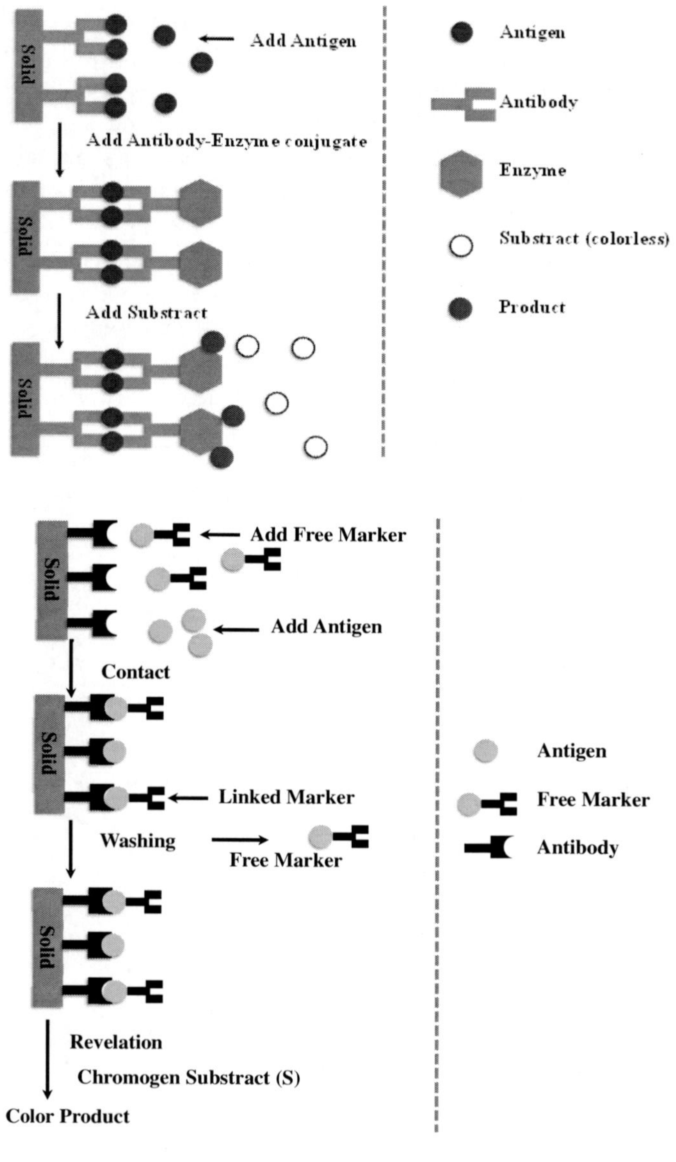

Figure 8. Schemes for possible immunosensors arrangement.

For the detection of antigens or antibodies in electrochemical systems two main approaches have been used (Figure 8), the first consists of a competitive type for detection of low molecular weight molecules; while the second, also called sandwich-type, like the first, is still a direct method but, in general, used to detect molecules of high mass.

The primary difference between them is that in direct competitive method of detection, the antibody is immobilized and reacts with the antigen which is in competition with the secondary antibody (enzyme-conjugated antibody) used as a marker (Ab*, of the structure Ab-Ag/Ab*) or immobilized antigen competes with the free antigen by labeled antibody type Ag:Ab*:Ag, considering the sandwich-type, the immobilized antigen competes with the free labeled antigen by the antibody, the last one is the most commonly employed immunosensors development.

The sandwich method comprises the following steps:

(a) Immobilization of antibody on the electrode surface;
(b) The antibody reacts against the antigen added in order to form the complex, which is represented by Ab-Ag;
(c) Another type of antibody (Ab*) is added to this complex, resulting in the formation of a final complex of structure Ab-Ag-Ab*.

As described, the complex formed (Ab-Ag-Ab*) must be electroactive, enabling the measurement signal through electroanalytical techniques, such as amperometric and voltammetric (Figure 9). Techniques are chosen according to the change in electrode surface Ag-Ab complex associated with other species physical, chemical, biological or electrochemical (Ricci et al., 2012).

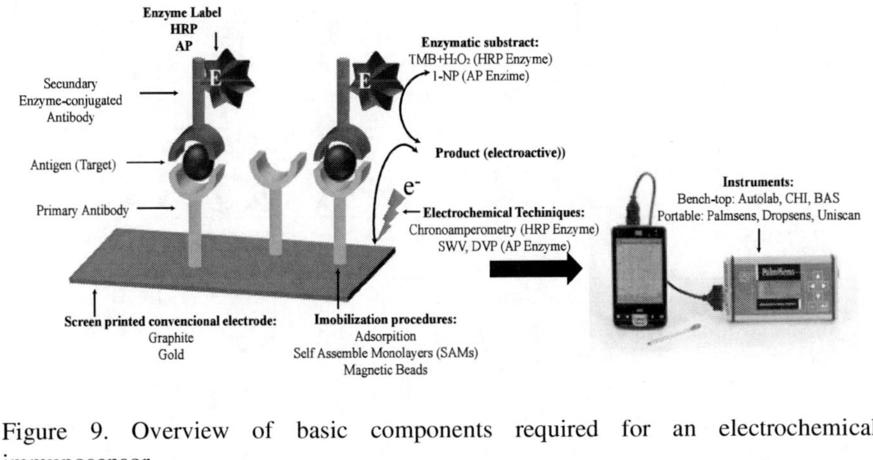

Figure 9. Overview of basic components required for an electrochemical immunosensor.

Currently, as well as for enzymatic biosensors, many electroanalytical techniques are already being used to capture the signal generated by the immunosensors, It is common to notice variations or significant dependence on pH and temperature for a few methods, however they are not considered in this section, since. The main electrochemical immunosensors are applied by different techniques

Over the last decade, various techniques exploiting imunossessors has been described, from potentiometry to optical imunnosensors, passing throw field-effect transistors and impedance spectroscopy electrochemistry, amperometry, voltammetry (linear sweep, cyclic, normal and differential pulse, square wave, and chronoamperometry) (Riccardi et al., 2002; D'Orazio, P., 2003; Holford et al., 2011).

Table 3. Some current and interesting examples of applications for immunosensors

Analyte	Technique, detection	Reference
Aflatoxins	Electrochemical impedance spectroscopy	Owino et al., 2007
Antibodies related to Chagas disease	Amperometric	Foguel, M. V, 2011
Brucella abortus CE-protein antigen	Amperometric	Gupra et al., 2014
Brucella bovis related proteins	Chemiluminescent	Liebes et al., 2009a Liebes et al., 2009b
Brucella melitensis	Electrochemical impedance spectroscopy	Wu et al., 2013
Carcinoembryonic antigen	Electrochemical	Ping et al., 2010
Escherichia coli O157:H7	Piezoelectric Electrochemical Fluorescence	Su et al., 2004 Radke et al., 2005 Yu, et al., 2002
Ebola virus	Chemiluminescent immunosensor	Petrosova et al., 2007
Hepatitis C virus	Piezoeletric	Skladal et al., 2004
Human chorionic gonadotrophin	Amperometric	Chai et al., 2008
Ochratoxin A	Electrochemical impedance spectroscopy	Yang et al., 2015
Prostate-specific antigen (PSA)	Amperometric	Healy et al., 2005

Besides their simplicity, immunoassays are often and cheaper when compared to other analytical methods such as chromatographies and spectroscopies, they do not need a more accurate treatment for samples, their time of analysis is significantly lower and the analytical response is of great qualitative and quantitative expression, furthermore in most cases, does not need technical expertise to perform.

Other advantage of using immunosensors is the possibility of automation, enabling specific tests in real time and in situ, and use a smaller amount of sample needed for detection of the pathogen (Rusling, 2010).

Despite all these pros, however, the development of new techniques using immunosensors still have limitations due antigens and antibodies are not electroactive molecules, requiring one element for recognition of the antigen or antibody on the electrode, and that the marking with a redox enzyme, the enzymatic reaction species produced becomes the final one electroactive molecule.

The wide range of applicability, illustrated in Table 3, is due to the above-mentioned aspects, which also and stimulates great efforts on improving immunosensors capabilities, including the detection of various analytes at the same time (so called multi-analyte detection), microfluidic and lab-on-a-chip systems, that promises to be decisive for the diagnosis of various important diseases, among others.

Molecularly Imprinted Sensors (MIP-based Sensors)

Molecularly imprinted polymers (MIPs) are synthetic materials with molecular recognition ability. These sorbents are synthesized by employing a template molecule that imprints its complementary physical-chemistry characteristics in the polymeric network (Sharma et al., 2013). After the synthesis, the template is then removed and the selective binding sites are, theoretically, free to recognize the template molecule during futures extractions.

The first MIP was synthesized in 1931 by Polyakov through the formation of selective cavities in silica by using a template (Polyakov et al., 1931). Since then other kind of supports were developed and used to prepare selective sorbents, as, for example, globulins and silica gel (Pauling, 1940, Dickey, 1949; Haldeman and Emmett, 1955; Bernhard, 1952), which were gradually replaced, in next years, by advantageous synthetic organic polymers (Wulff and Sarhan, 1972; Sagiv, 1980; Cohen et al. 1986). Currently, the best alternative for molecularly imprinting technology is based on organic polymers (Dias et al., 2008), frequently synthesized by using at least four important reagents (see Figure 10): a-) functional monomers, b-) template molecule, c-) cross linker, and d-) initiator.

All the reagents are dissolved in a suitable solvent, and the reaction is then initiated by using high temperatures or UV radiation (depending of the initiator nature). Functional monomers need to be chemically complementary to the template. As example, the use of acidic functional monomers is requested to prepare a MIP selective to the weak basis, and vice-versa. The nature of the interactions needs to be preferably non-covalent, in order to use elution mechanism based on organic solvents, ionic strength, and pH. The functional monomer also needs to present polymerizable groups (e.g. vinyl group) able to react with the cross-linker, promoting the interconnection between the polymeric chains. The cross-linker can be bi- or trifunctional, depending of the requested density of the polymeric network (Dias et al., 2008). Additionally, the amount of cross-linker influences directly in the flexibility of the polymer, resulting in rigid or flexible binding sites for high and low cross-linker concentrations, respectively.

The initiator is used to generate the first radical to be transferred to the functional monomer or cross-linker, starting a chain reaction.

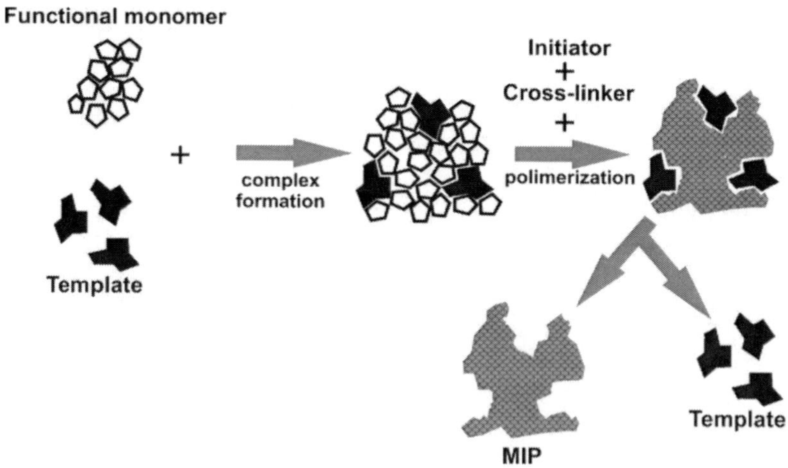

Figure 10. Scheme of the MIP synthesis.

Table 4. Some current and interesting examples of applications for MIP-based sensors

Analyte	Technique, detection	Reference
Bisphenol A	Voltammetric	Huang et al., 2011
Caffeine	Voltammetric	Alizadeh et al., 2010 (a)
Estradiol	Voltammetric	Azevedo et al., 2013
Parathion	Voltammetric and microgravimetric	Marx et al., 2004
Promethazine	Voltammetric	Alizadeh et al., 2012
Testosterone	Electrochemical impedance spectroscopy	Betatache et al., 2014
Trinitrotoluene	Voltammetric	Alizadeh et al., 2010 (b)

The MIPs have been extensively used in several applications with emphasis on chemical sensing (Blanco-Lopez et al., 2004; Stephenson and Shimizu, 2007) and sample preparation procedures (Martin-Esteban, 2013).

Particularly in chemical sensors, MIPs are selective units coupled onto transducer surfaces that converts the amount of bound molecules in a measurable analytical signal. In this case, MIPs have the same role of biomolecules (e.g. enzymes, antibodies) in biosensors, with many advantages regarding costs, chemically and physically stability, ease of prepare-handle-stock-transport, among others.

An important point in sensors with MIPs is the process used to fix the polymer onto transducing surface (Tarley et al., 2005). In cases when the sensor is a solid electrode, the immobilization can be carried out by using a mixture of MIP and a supporting material that is applied directly on the transducer surface; as well as by direct formation of a MIP film *in situ*. In the first case, the main example is the mixed of MIP with carbon-based powder (graphite or graphene) resulting in a composite that is physically immobilized in a support (Zaidi and Shin, 2014). In this kind of sensors, the interface can be easily renewed for each extraction, by a simple polishing step.

The strategies for *in situ* preparation of MIP films are the most described in literature and involve several different procedures (Sharma et al., 2013), including: a-) electropolymerization employing an electroactive functional monomer (Sharma et al., 2012), b-) drop-casting of a polymerization solution (Holthoff and Bright, 2007), c-) preparation of membranes from a PVC binder containing the MIP and a conducting material as carbon nanotubes, graphite, and graphene (Diaz-Diaz et al., 2012), d-) *in situ* chemical polymerization by adding the polymerizable solution on the support in spin, forming a thin film (Guan et al., 2011), and e-) self-assembled of monolayers (Balamurugan and Spivak, 2011).

Based on the wide range of advantages and perspectives, many applications for such technology can be found in literature, among them we highlight some of interesting and current approach (Table 4):

MIP-based sensors are in general, by means of sensitivity comparison, still inferior to biosensors. On the other hand, the great stability, low costs, and possibility of modeling different polymers based on many different templates; including molecules without natural biochemical recognition are still advantages that encourage scientists. Most of these applications are dependent on the related mass and charge transport properties. Thus, it becomes essential to understand the mechanisms of those transports in these materials, what can be carried out by electrochemical impedance spectroscopy, applicable for characterization, modelling and understanding interfacial processes involving the mass and charge transport on surfaces of modified electrodes. The analytical signal generated by the above-mentioned interacting systems can be translated by various electroanalytical methods, specially voltammetry and amperometry, being their main advantages the possibility of analyzing *in situ* smaller amounts of sample, in a shorter time.

ELECTRONIC TONGUES AND NOSES: PROMISING TRENDS

Based on all above-cited clues some trends regarding sensing technologies could be easily pointed. Different platforms, initially used as direct sensors, are capable of supporting chemical modification, specially onto their surface, not only by mixing components, but specially by chemisorption, electrodeposition or electropolymerization of different molecules, mediators and probes, metal and oxide thin films deposition, among others; it is certainly an important way of improve sensor´s properties. Such modifications allow a wide range of potential uses, from deposition of enzymes, antibodies, deoxyribonucleic acids to the use o MIP´s and other synthetic recognition mechanisms.

All these options provide more selective and sensitive sensors (enzymatic biosensors, immunosensors, or even multiplexes systems) capable of recognize, and quantify specific components in complex matrices. In addition, miniaturized customized devices take special place in field and/or *in situ* analysis, as well as the development of electronic tongues and noses. In this context, sensing and monitoring fields are being constantly affected by new materials, being nanotechnology and materials science special cores of such promising future.

ACKNOWLEDGMENTS

Authors are grateful to Coordenação de Aperfeiçoamento de Pessoal de Nível Superior (CAPES), Conselho Nacional de Desenvolvimento Científico e Tecnológico (CNPq, process 483371/2012-2), Fundação de Amparo a Pesquisa do Estado de Minas Gerais (FAPEMIG, process CEX - APQ-01556-13), Fundação Carlos Chagas de Amparo a Pesquisa do Estado do Rio de Janeiro (FAPERJ, processes E-26/210.055/2014 and E-26/102.971/2012), Fundação de Amparo a Pesquisa do Estado de São Paulo (FAPESP), and Pró-reitoria de Pesquisa, Pós-graduação e Inovação (Proppi-UFF), for constant financial support.

REFERENCES

Adams, R. N. (1958). Carbon Paste Electrodes. *Analytical Chemistry*, *10*, 9, 1576.

Adams, R. N. (1969). Electrochemistry at solid electrodes. Dekker: New York.

Agostinho, S. M. L., Villamil, R. F. V., Neto, A. A. & Aranha, H. (2004). The supporting electrolyte and its multiple functions on electrode processes. *Química Nova*, *27*, 5, 813-817.

Alizadeh, T. & Ganjali, M. R. (2012). Akhoundian, M. Fabrication of an extra sensitive voltammetric sensor using nanoparticles of molecularly imprinted polymer for determination of ultra-trace promethazine in plasma sample. *International Journal of Electrochemical Sciense*, *7*, 10427-10441.

Alizadeh, T., Ganjali, M. R., Zare, M. & Norouzi, P. (2010). Development of a voltammetric sensor based on a molecularly imprinted polymer (MIP) for caffeine measurement. *Electrochimica Acta*, *55*, 1568-1574.

Alizadeh, T., Zare, M., Ganjali, M. R., Norouzi, P. & Tavana, B. (2010). A new molecularly imprinted polymer (MIP)-based electrochemical sensor for monitoring 2,4,6-trinitrotoluene (TNT) in natural waters and soil samples. *Biosensors and Bioelectronics*, *25*, 1166-1172.

Ansari, S. A. & Husain, Q. (2012). Potential applications of enzymes immobilized on/in nano materials: A review. *Biotechnology Advanves*, *30*, 512-523.

Arredondo Valdez, H. C., García Jiménez, G., Gutiérrez Granados, S. & Ponce de León, C. (2012). Degradation of paracetamol by advance oxidation processes using modified reticulated vitreous carbon electrodes with TiO_2 and $CuO/TiO_2/Al_2O_3$. *Chemosphere*, *89*, 10, 1195-1201.

Azevedo, S., Lakshmi, D., Chianella, I., Whitcombe, M. J., Karim, K., Ivanova-Mitseva, P. K., Subrahmanyam, S. & Piletsky, S. A. (2013). Molecularly imprinted polymer-hybrid electrochemical sensor for the detection of β-estradiol. *Industrial & Engineering Chemistry Research*, *52*, 13917-13923.

Bagotsky, V. S. (2006). Fundamentals of electrochemistry. John Wiley & Sons.

Balamurugan, S. & Spivak, D. A. (2011). Molecular imprinting in monolayer surfaces (Review). *Journal of Molecular Recognition*, *24*, 6, 915-929.

Bard, A. J. & Faulkner, L. R. (1980). Electrochemical Methods: Fundamentals and Applications. Wiley.

Barek, J. (2002). Eighty Years of Polarography: History and Future. *Electroanalysis*, *15*, 5-6, 67-472.

Bernhard, S. A. (1952). The preparation of specific adsorbants. *Journal of the American Chemical Society*, *74*, 4946-4947.

Betatache, A., Lagarde, F., Sanglar, C., Bonhomme, A., Leonard, D. & Jaffrezic-Renault, N. (2014). Gold electrodes modified with molecular imprinted acrylate polymer for impedimetric determination of testosterone. *Sensors & Transducers*, *27*, 92-99.

Blanco-Lopez, M. C., Lobo-Castanon, M. J., Miranda-Ordieres, A. J. & Tunon-Blanco, P. (2004). Electrochemical sensors based on molecularly imprinted polymers. *Trends in Analytical Chemistry*, *23*, 1, 36-48.

Bockris, J. O. M. & Reddy, A. K. N. (1970). Modern electrochemistry, an introduction to an interdisciplinary area. Plenum Press: New York.

Brett, C. M. A. & Brett, A. M. O. (1993). Electrochemistry : principles, methods and applications. Oxford University Press: Oxford.

Cervini, P., Claro-Neto, S. & Cavalheiro, E. T. G. (2002). Evaluation of a new rigid carbon-castor oil polyurethane composite as an electrode material. *Talanta*, *57*, 909-917.

Chai, R., Ou, C., Yuan, R., Chay, Y., Caos, S. & Li, X. (2008). Amperometric immunosensors based on layer-by-layer assembly of gold nanoparticles and methylene blue on thiourea modified glassy carbon electrode for determinations of human chorionic gonadotrophin. *Talanta*, *74*, 1330-1336.

Chaubey, A. & Malhotra, B. D. (2002). Mediated biosensors. *Biosensors & Bioeletronics*, *17*, 441-456.

Chen, P. & McCreery, R. L. (1996). Control of electron transfer kinetics at glassy carbon electrodes by specific surface modification. *Analytical Chemistry*, *68*, 3958-3965.

Cipolatti, E. P., Silva, M. J. A., Klein, M., Feddern, V., Feltes, M. M. C., Oliveira, J. V., Ninow, J. L. & Oliveira, D. (2014). Currents status and trends in enzymatic nanoimmobilization. *Journal of Molecular Catalysis B: Enzymatic*, *99*, 56-67.

Ciriello, R., Cataldi, T. R., Crispo, F. & Guerrieri, (2015) A. Quantification of L-lysine in cheese by a novel amperometric biosensor. *Food Chemistry*, *169*, 13-19.

Cohen, S. R., Naaman, R. & Sagiv, J. (1986). Thermally induced disorder in organized organic monolayers on solid substrates. *Journal of Physical Chemistry*, *90*, 3054-3056.

Compton, R. G., Banks, C. E., Editors, (2007). Understanding Voltammetry. World Scientific.

Cruys-Bagger, N., Ren, G., Tatsumi, H., Baumann, M. J., Spodsberg, N., Andersen, H. D., Gorton, L., Borch, K. & Westh, P. (2012). An amperometric enzyme biosensor for real-time measurements of cellobiohydrolase activity on insoluble cellulose. *Biotechnology and Bioengineering*, *109*, 12, 3199-3204.

Csöregi, E., Gorton, L. & Marko-Varga, G. (1993). Carbon fibres as electrode materials for the construction of peroxidase-modified amperometric biosensors. *Analytica Chimica Acta*, *273*, 1-2, 59-70.

Demirkiran, N., Ekinci, E. & Asiltürk, M. (2012). Immobilization of glucose oxidase in silica sol-gel film for application to biosensor and amperometric determination of glucose. *Journal of the Chilean Chemical Society*, *57*, 4, 1336-1339.

Dias, A. C. B., Figueiredo, E. C., Grassi, V., Zagatto, E.A.G. & Arruda, M. A. Z. (2008). Molecularly imprinted polymer as a solid phase extractor in flow analysis. *Talanta*, *76*, 988-996.

Diaz-Diaz, G., Antuna-Jimenez, D., Blanco-Lopez, M. C., Lobo-Castanon, M. J., Miranda-Ordieres, A. J. & Tunon-Blanco, P. (2012). New materials for analytical biomimetic assays based on affinity and catalytic receptors prepared by molecular imprinting. *Trends in Analytical Chemistry*, *33*, 68-80.

Dichey, F. H. (1949). The preparation of specific adsorbents. *Proc. Natl. Acad. Sci.*, *35*, 5, 227-229.

D'Orazio, P. (2003). Biosensors in clinical chemistry. *Clinica Chimica Acta*, *334*, 2, 41-69.

Edmonds, T. E. (1985). Electroanalytical applications of carbon fibre electrodes. *Analytica Chimica Acta*, *175*, 0, 1-22.

Ekabutr, P., Chailapakul, O. & Supaphol, P. (2013). Modification of disposable screen-printed carbon electrode surfaces with conductive electrospun nanofibers for biosensor applications. *Journal of Applied Polymer Science*, *130*, 6, 3885-3893.

Engstrom, R. C. (1982). Electrochemical pretreatment of glassy carbon electrodes. *Analytical Chemistry*, *54*, 2310-14.

Engstrom, R. C. & Strasser, V. A. Characterization of electrochemically pretreated glassy carbon electrodes. *Analytical Chemistry*, *56*, 136-41.

Feng, J. X., Brazell, M., Renner, K., Kasser, R. & Adams, R. N. (1987). Electrochemical pretreatment of carbon fibers for in vivo electrochemistry: effects on sensitivity and response time. *Analytical Chemistry*, *59*, 14, 1863-1867.

Fisher, W. L. B., & Kelley, M. T. (1966). In "Polarography, 1964". Ed. by, Hills, G. J., Polarographic Society (Great Britain), Polarography 1964: proceedings of the Third International Congress, Southampton. Macmillan: London.

Foguel, M. V. (2011). Amperometric immunosensor for chagas disease using gold CD-R transducer. *Electroanalysis*, *23*, 2555-2561.

Gerard, M., Chaubey, A. & Malhotra, B. D. (2002). Application of conducting polymers to biosensors. *Biosensors and Bioeletronics*, *17*, 5, 345-359.

Guan, G., Wang, S., Zhou, H., Zhang, K., Liu, R., Mei, Q., Wang, S. & Zhang, Z. (2011). Molecularly imprinted polypyrrole nanonecklaces for detection of herbicide through molecular recognition-amplifying current response. *Analytica Chimica Acta*, *702*, 2, 239-246.

Gupra, A. K., Rao, V. K., Selvam, D. T. Kumar, A. & Jain, R. (2014). Amperometric immunosensor of *Brucella abortus* CE-protein antigen shows post-zone phenomena. *Journal of Electroanalytical Chemistry*, *717*, 83-89.

Haldeman, R. G. & Emmett, P. H. (1955). Specific adsorption of alkyl orange dyes on silica gel. *Journal of Physical Chemistry*, *59*, 1039-1043.

Healy, D. A., Hayes, C. J., Leonard, P., Mckenna, L. & O' Kennedy, R. (2007). Biossensor developments: application to prostate-specific antigen detection. *Trends in Biotechnology*, *25*, 125-131.

Hocevar, S. B., Svancara, I., Vytras, K. & Ogorevc, B. (2005). Novel electrode for electrochemical stripping analysis based on carbon paste modified with bismuth powder. *Electrochimica Acta*, *51*, 706-710.

Holford, T. R., Davis, F. & Higson, S. P. (2011). Recent trends in antibody based sensors. *Biosensors and Bioelectronics*, *34*, 1, 12-24.

Holthoff, E. L., Bright, F. V. (2007). Molecularly templated materials in chemical sensing. *Analytica Chimica Acta*, *594*, 2, 147-161.

Huang, J., Zhang, X., Lin, Q., He, X., Xing, X., Huai, H., Lian, W. & Zhu, H. (2011). Electrochemical sensor based on imprinted solegel and nanomaterials for sensitive determination of bisphenol A. *Food Control*, *22*, 786-791.

Ito, T., Shimomura, T., Hayashi, A., Yamaguchi, A., Teramae, N., Ono, M., Tsunoda, T., Mizukami, F., Stucky, G. D. & Hanaoka, T. A. (2014). Electrochemical enzymatic biosensor with long-term stability using hybrid mesoporous membrane. *Analyst*, *139*, 18, 4654-4660.

Johnson, D. C. & LaCourse, W. R. (1990). Liquid chromatography with pulsed electrochemical detection at gold and platinum electrodes. *Anaytical Chemistry, 62*, 589-597.

Kalcher, K., Kauffmann, J. M., Wang, J., Švancara, I., Vytřas, K., Neuhold, C. & Yang, Z. (1995). Sensors based on carbon paste in electrochemical analysis: A review with particular emphasis on the period 1990–1993. *Electroanalysis*, *7*, 1, 5-22.

Kiliç, E. & Aslan, N. (2005). Determination of autoprotolysis constants of water-organic solvent mixturesby potentiometry. *Microchimica Acta, 151*, 89-92.

Krajewska, B. (2004). Application of chitin- and chitosan-based materials for enzyme immobilizations: a review. *Enzyme and Microbial Technology*, *35*, 126-139.

Kutner, W., Wang, J., L'Her, M. & Buck, R. P. (1998). Analytical aspects of chemically modified electrodes: classification, critical evaluation and recommendations. *Pure & Applied Chemistry*, *70*, 6, 1301-1318.

Li, C., Wu, Z., Hartings, J.A., Rajan, N., Chahine, N., Cheyuo, C., Wang, P., Wu, P. M., Golanov, E.V., Ahn, C. H._& Narayan, R. K. Brain-friendly amperometric enzyme biosensor based on encapsulated oxygen generating biomaterial, *Proceedings of the Annual International Conference of the IEEE Engineering in Medicine and Biology Society, EMBS* 2012, Pages 6003-6006. 34th Annual International Conference of the IEEE Engineering in Medicine and Biology Society, United States.

Li, Z., Cong, F., Shen, G. & Yu, R. (2002). Bacteria-modified Amperometric Immunosensor for a *Brucella melitensis* Antibody Assay. *The Japan Society for Analytical Chemistry*, *18*, 625-629.

Liebes, Y., Amir, L., Marks, R. S. & Banai, M. (2009a). Immobilization strategies of Brucella particles on optical fibers for use in chemiluminescence immunosensors. *Talanta*, *80*, 338-345.

Liebes, Y., Marks, R. S. & Banai, M. (2009b). Chemiluminescent optical fiber immunosensor detection of Brucella cells presenting smooth-A antigen. *Sensors and Actuators B*, *140*, 568-576.

Liu, Q., Hu, N., Zhang, F., Wang, H., Ye, W. & Wang, P. (2012). Neurosecretory cell-based biossensor: monitoring secretion of adrenal chromaffin cells by local extracellular acidification using light-addressable potenciometric sensor. *Biosensors and Bioeletronics*, *35*, 1, 421-424.

Liu, X., Lin, P., Yan, X., Kang, Z., Zhao, Y., Lei, Y., Li, C., Du, H. (2009). Zhang, Y. Enzyme-coated single ZnO nanowire FET biossensor prepared by aptamer-assisted immobilitation. *Sensors and Actuators B: Chemical*, 2012. Inpress. DOI: 10.1016/jsnb 2013.02.

Malhotra, B. D., Chaubey, A. & Singh, S. P. (2006). Prospects of conducting polymers in biosensors. *Analytica Chimica Acta*, *578*, 1, 59-74.

Marx, S., Zaltsman, A., Turyan, I. & Mandler, D. (2004). Parathion sensor based on molecularly imprinted sol-gel films. *Analytical Chemistry*, *76*, 120-126.

Martin-Esteban, A. (2013). Molecularly-imprinted polymers as a versatile, highly selective tool in sample preparation *Trends in Analytical Chemistry*, *45*, 169-181.

McCreery, R. L. "Carbon electrodes: Structural effects on electron transfer kinetics," in Bard, A. J. (1991). Electroanalytical Chemistry, Marcel Dekker, New York.

Navratilova, Z. & Kula, P. (2000). Study of ion exchange on vermiculite by means of carbon paste electrode. *Scr.--Geol.*, 28/29, 43-54.

Nlirnberg, H. W. (1979). Polarography and voltammetry in studies of toxic metals in man and his environment. *The Science of the Total Environment*, *12*, 35-60.

Oliveira, A. C., Santos, S. X. & Cavalheiro, E. T. G. (2008). Graphite–silicone rubber composite electrode: Preparation and possibilities of analytical application. *Talanta*, *74*, 4, 1043-1049.

Owino, J. H. O., Ignaszak, A., Al-Ahmed, A., Baker, P. G. L., Alemu, H., Ngila, J. C. & Iwuoha, E. I. (2007). Modelling of the impedimetric responses of an aflatoxin B1 immunosensor prepared on an electrosynthetic polyaniline platform. *Analytical and Bioanalytical Chemistry*, *388*, 1069.

Pauling, L. J. (1940). A Theory of the structure and process of formation of antibodies. *Journal American Chemical Society*, *62*, 10, 2643-2657.

Pauliukaite, R., Ghica, M. E., Fatibello-Filho, O. & Brett, C. M. A. (2009). Comparative study of different cross-linking agents for the immobilization of functionalized carbon nanotubes within a chitosan film supported on a graphite–epoxy composite electrode. *Analytical Chemistry*, *81*, 13, 5364-5372.

Perantoni, C. B., Carbogim, L. G. S., Semaan, F. S., Matos R. C. & Lowinsohn, D. (2011). Flow injection analysis of ethambutol in antituberculosis Drugs using a graphite-paraffin electrode as amperometric detector. *Electroanalysis*, *23*, 11, 2582-2585.

Pereira, A. C., Santos, A. S. & Kubota, L. T. (2002). Trends in amperometric electrodes modification for electroanalytical applications. *Química Nova*, *25*, 6, 1012-1021.

Petrosova, A., Konry, T., Cosnier, S., Trakht, I., Lutwama, J., Rwaguma, E., Chepurnov, A., Muhlberger, E.,Lobel, L. & Marks, R. S. (2007). Development of a highly sensitive, field operable biosensor for serological studies of Ebola virus in central Africa. *Sensors and Actuators B*, *122*, 578-586.

Ping, L., Ligen, M., Ruo, Y., Yaquin, C. & Shihong, C. (2010). A novel immunosensor for carcinoembryonic antigen based on poly(diallyldimethylammonium chloride) protect prussian blue nanoparticles and double-layer nanometer-sized gold particles. *Microchimica Acta*, *1713*, 297-304.

Polyakov, M. V. (1931). Adsorption proprieties and structure of the silica gel. *Zhurnal Fizicheskoi Khimii*, *2*, 799-805.

Radke, S. M., Alocilja, E. C. (2005). A high density microelectrode array biosensor for detection of E. coli O157:H7. *Biosensors and Bioelectronics*. *20*, 1662-1667.

Riccardi, C. S., da Costa, P. I. & Yamanaka, H. (2002). Amperometric immunosensor. *Química Nova*, *25*, 2, 316-320.

Ricci, F., Adormetto, G. & Palleschi, G. (2012). A review of experimental aspects of electrochemical immunosensors. *Eletrochemica Acta*, *84*, 74-83.

Rice, M. E., Galus, Z. & Adams, R. N. (1983). Graphite paste electrodes: effects of paste composition and surface states on electron-transfer rates.

Journal of Electroanalytical Chemistry and Interfacial Electrochemistry, 143, 1-2, 89-102.

Roberts, M., Huang, A. F., Johns, P. & Owen, J. (2013). Dip–spin coating of reticulated vitreous carbon with composite materials to act as an electrode for 3D microstructured lithium ion batteries. *Journal of Power Sources, 224*, 0, 250-259.

Rusling, J. F., Kumar, C. V., Gutkind, J. S. & Patel, V. (2010). Measurement of biomarker proteins for point-of-care early detection and monitoring of cancer. *Analyst, 135*, 10, 2496-2511.

Sagiv, J. (1980). Organized monolayers by adsorption I. Formation and structure of oleophobic mixed monolayers on solid surface. *Journal of the American Chemical Society, 102*, 92-98.

Santos, T. A. D., Barreto, L. M., Ritta, A. G. S. L., Meneses, W. S., Nunes, R. S. & Semaan, F. S. (2013). Cost-effective composite electrode for the Fast voltammetric screening and determination of riboflavin (B2) and pyridoxine (B6) in pharmaceuticals. *Revista Virtual de Química, 5*, 4, 548-562.

Sassolas, A., Blum, J. L. & Leca-Bouvier, B. D. (2012). Immobilization strategies to develop enzymatic biosensors. *Biotechnology Advances, 30*, 489-511.

Sawyer, D. T., Sobkowiak, A. & Roberts, J. L. (1995). Electrochemistry for chemists. 2. ed., Wiley: New York.

Scholz, F., Editor, (2005). Electroanalytical Methods: Guide to Experiments and Applications, 2nd printing. Springer.

Sharma, P. S., Dabrowski, M., D’Souza, F. & Kutner, W. (2013). Surface development of molecularly imprinted polymer films to enhance sensing signals. *Trends in Analytical Chemistry, 51*, 146-157.

Sharma, P. S., Pietrzyk-Le, A., D'Souza, F. & Kutner, W. (2012). Electrochemically synthesized polymers in molecular imprinting for chemical sensing. *Analytical and Bioanalytical Chemistry, 402*, 10, 3177-3204.

Sharma, S. K., Sehgal, N. & Kumar, A. (2003). Biomolecules for development of biosensors and their applications. *Current Applied Physics, 3*, 307-316.

Shi, X., Gu, W., Li, B., Chen. N., Zhao, K. & Xian, Y. (2014). Enzymatic biosensors based on the use of metal oxide nanoparticles. *Microchimica Acta, 181*, 1-22.

Skladal, P., Riccardi, C. S., Yamanaka, H. & da Costa, P. I. (2004). Piezoelectric biosensors for real-time monitoring of hybridization and detection of hepatitis C virus. *Journal of Virological Methods*, *117*, 2, 145-151.

Souza, M. F. B. (1997). Chemically modified electrodes applyed to electroanalysis: a brief presentation. *Química Nova*, *2*, 2, 191-195.

Sowmya, A. (2014). Meenakshi, S. A novel quaternized chitosan–melamine–glutaraldehyde resin for theremoval of nitrate and phosphate anions. *International Journal of Biological Macromolecules*, *64*, 224–232.

Stephenson, C. J. & Shimizu, K. D. (2007). Colorimetric and fluorometric molecularly imprinted polymer sensors and binding assays. *Polymer International*, *56*, 4, 482-488.

Su, X. L. & Li, Y. (2004). A self-assembled monolayer-based piezoelectric immunosensor for rapid detection of Escherichia coli O157:H7. *Biosensors and Bioelectronics*. *19*, 563-574.

Svancara, I., Vytrias, K., Barek, J. & Zima, J. (2010). Carbon paste electrodes in modern electroanalysis. *Critical reviews in Analytical Chemistry*, *31*, 4, 311-345.

Taleat, Z., Khoshroo, A. & Mazloum-Arkadani, M. (2014). Screen-printed electrodes for biosensing: a review (2008–2013). *Microchimica Acta*, *181*, 865-891.

Tallman, D. E., Petersen, S. L. (1999). Composite electrodes for electroanalysis: Principles and applications. *Electroanalysis*, *2*, 499-510.

Tarley, C. R. T., Sotomayor, M. D. T. & Kubota, L. T. (2005). Biomimetic polymers in analytical chemistry. Part 2: applications of MIP (Molecularly Imprinted Polymers) in the development of chemical sensors. *Química Nova*, *28*, 6, 1087-1101.

Thevenot, D. R. et al. (2001). Eletrochemical biosensors: recommended definitions and classifications. *Biosensors and Bioelectronics*, *16*, 121-131.

Turner, A. P. F. (2013). Biosensors: sense and sensibility. *Chemical Society Reviews*, *42*, 8, 3175-3648.

Wu, H., Zuo, Y., Cui, C., Yang, W., Ma, H. & Wang, X. (2013). Rapid quantitative detection of *Brucella melitensis* by a label-free impedance immunosensor based on a gold nanoparticle-modified screen-printed carbon electrode. *Sensors*, *12*, 8551-8563.

Wulff, G. & Sarhan, A. (1972). Use of polymers with enzyme-analogous structures for the resolution of racemates. *Angewandte Chemie, International Edition*, *11*, 341-344.

Yang, H. (2012). Enzyme-based ultrasensitive electrochemical biosensors. *Current Opinion in Chemical Biology*, *16*, 422-428.

Yang, J., Gao, P., Liu, Y., Li, R., Ma, H., Du, B. & Wei, Q. (2015). Label-free photoelectrochemical immunosensor for sensitive detection of Ochratoxin A. *Biosensors and Bioelectronics*, *64*, 13-18.

Yu, L. S. L., Reed, S. A. & Golden, M. H. (2002). Time-resolved fluorescence immunoassay for the detection of Escherichia coli O157:H7 in apple cider. *Journal of Microbiological Methods*. *49*, 63-68.

Zaidi, S. A. & Shin, J. H. (2014). Molecularly imprinted polymer electrochemical sensors based on synergistic effect of composites synthesized from graphene and other nanosystems. *International Journal of Electrochemical Science*, *9*, 8, 4598-4616.

In: New Developments in Analytical Chemistry ... ISBN: 978-1-63463-427-4
Editor: Breanna Granger

Chapter 3

ENHANCED ELECTROCHEMICAL PROPERTIES OF MULTI-WALLED CARBON NANOTUBE PASTE ELECTRODE FOR SIMULTANEOUS DETERMINATION OF PHENOLIC COMPOUNDS IN THE PRESENCE OF SURFACTANT

Felipe Augusto Gorla[1], Eduardo Henrique Duarte[1], Juliana de Fátima Giarola[2], Elen Romão Sartori[1], Arnaldo César Pereira[2] and César Ricardo Teixeira Tarley*[*,1,3]

[1]Universidade Estadual de Londrina (UEL), Departamento de Química, Londrina, PR, Brazil
[2]Universidade de São João del Rei (UFSJ), Departamento de Ciências Naturais, São João del-Rei, MG, Brazil
[3]Departamento de Química Analítica - Instituto Nacional de Ciência e Tecnologia (INCT) de Bioanalítica, Universidade Estadual de Campinas (UNICAMP), Cidade Universitária Zeferino Vaz s/n, Campinas, Brazil

[*] Corresponding author. Tel +55 43 3371 4366; Fax +55 43 3371 4286; E-mail address:ctarleyquim@yahoo.com.br (C. R. T. Tarley).

ABSTRACT

Carbon-based nanomaterials such as carbon nanotubes (CNT), carbon black and fullerenes are interesting materials for the development of analytical methods. In electrochemistry carbon nanotubes improves the electron transfer and provides an improvement in the analytical sensitivity and selectivity of phenolic compounds determination. Furthermore, the electroanalytical behavior of CNT toward phenolic compounds can be increased using measures in the presence of surfactants. In this way, this work deals with study of a voltammetric method for simultaneous determination of four emerging pollutants Hydroquinone (HQ), Catechol (CT), Guaiacol (GUA) and 4-Nitrophenol (pNF) using a carbon nanotubes paste sensor (37% of CNT and 63% w/w of mineral oil) whose measures were performed in presence of a cationic surfactant (cetylpyridiniumbromide, CPB). The oxidation peaks for the four phenolic compounds were overlapped in the absence of CPB, while a significant increase on oxidation peak current, a shift of potential towards less positive values and a satisfactory separation of oxidation potential in presence of CPB (0.45 mmol L^{-1}) allowed the simultaneous determination. According to relationship of pH and oxidation potential it was observed a Nerstian system where the number of protons equal to the electrons. From chronoamperometric studies and using Cottrell equation the average diffusion coefficients were found to be 5.45×10^{-6}, 3.35×10^{-5}, 2.94×10^{-4}, 3.22×10^{-4} cm^2 s^{-1}, respectively for HQ, CT, GUA and pNF. Under optimized condition, i.e. pH 8.0, BR buffer at 0.1 mol L^{-1} and CPB at 0.45 mmol L^{-1} and linear sweep voltammetry at scan rate of 120 mV s^{-1}, limits of detection (LOD) and quantification (LOQ) in μmol L^{-1} were found to be 3.43 and 11.43 (HQ), 1.72 and 5.73 (CT), 1.10 and 3.67 (GUA) and 4.96 and 16.53 (pNF). The method has been applied for simultaneous determination of these phenolic compounds in water samples.

Keywords: Emerging pollutants, cyclic voltammetry, carbon-based nanomaterials, cetylpyridiniumbromide

1. INTRODUCTION

Emerging pollutants belongs to a new class of contaminants usually found at low levels (μg L^{-1} or ng L^{-1}) in water, soil and air (Urtiagaet al.,2013). The emerging pollutants are substances, in general of organic nature, that are not essentially new, but that are being found especially in wastewater and aquatic environments due to increasing of world population and agricultural and industrial practices, whose effects on environment and human health are unknown. Moreover, due to lack of knowledge of their impact in the middle or long-term effect on human health, there is no specific regulation to these substances for establishing the maximum level allowed in water samples (Murray et al., 2010). Some emerging pollutants can be cited such as pharmaceuticals, personal care products, flame retardants, plasticizers, organophosphorus compounds, metabolites of pesticides, degradation byproducts, among other (Barcelo, 2003; Eggenet al., 2010; Muñoz et al., 2008).

Regarding the degradation byproducts, the phenolic compounds have received considerable interest due to their toxicity and persistence in the environment. Some phenolic compounds have been already elucidated in scientific literature and regularized by Environmental Protection Agencies, such as the Brazilian National Council for the Environment (CONAMA 2014), which establishes a maximum amount of 0.5 mg L^{-1} of phenolic compounds for effluent discharges in aquatic bodies, U.S.EPA (U.S. Environmental Protection Agency) and European Commission for Environment (ECE) (Silva & Collins, 2011); however, some phenolic compounds are not regularized including the hydroquine, cathecol, guaiacol and 4-nitrophenol (Figure 1).

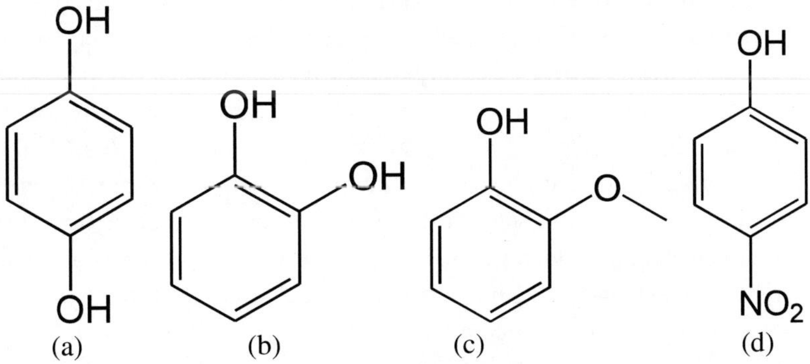

Figure 1. Molecular structure of hydroquine (a), cathecol (b), guaiacol (c) and 4-nitrophenol (d).

Phenolic compounds can be determined by UV-vis spectrophotometry, high performance liquid chromatography and gas chromatography (Rosattoet al., 2011). The spectrophotometric techniques presents low cost for acquisition and ease operation; however, suffers from selectivity for phenolic compounds.

Regarding the chromatographic ones using spectrophotometric and electrochemical detection, the phenolic compounds determination is widely performed (Bussyet al., 2012). Nevertheless, the acquisition of chromatographic system is very expensive and the sample throughput is very high. On the other hand, the electrochemical techniques are pratically attractive, due to high sensitivity, low cost for acquisition and easy portability. Despite these features, the simultaneous determination of phenolic compounds using conventional electrodes such as glassy carbon, carbon paste electrode and metallic electrodes (silver, gold, platinum and mercury) is very difficult owing to the overlapping of oxidation or reduction peaks. Thus, it is very important to research new electrodic materials for sensing phenolic compounds and different analytes. In this sense, the importance of nanomaterial chemistry is nowadays uncontested and new and sophisticated materials with surprising properties have promoted the approximation between chemistry and material sciences. Among the materials classified as nanomaterials, carbon nanotubes (CNTs) are the most representative. CNTs can be described as a graphite sheet rolled up into a nanoscale-tube (which are single-wall carbon nanotubes (SWCNTs), or with additional graphene tubes around the core of an SWCNT (which are multi-wall CNTs (MWCNTs). These CNTs have diameters in the range between fractions of nanometers and tens of nanometers and lengths up to several centimeters with both their ends normally capped by fullerene-like structures. The combination of size, structure and topology gives nanotubes important mechanical and surface properties. The electrical properties of CNTs depend sensitively on their diameter and chirality. According to the structural parameters, SWCNTs can be either a metal, semiconductor or small-gap semiconductor. Since the topological defects in nanotubes result in local perturbations of their electronic structure, the pentagonal defect of the caps make them more metallic than the cylinders. Due to their interesting properties, such as high surface area, high chemical and physical resistance and electrical properties, great attention has been paid in recent years in the development of new electroanalytical methods covering all the field of analytical science (Afkhamiet al., 2014, Karadas&Ozkan, 2014, Madrakianet al., 2014, Gaoet al., 2014, Wang &Arash, 2014, Yuan et al., 2014, Wei et al., 2014).

As example, Yin et al., 2011 used as a strategy for the peak separation of catechol, hydroquinone and resorcinol, the modification of glassy carbon surface with chitosan/graphene composite. Prathapet al., 2013 using glassy carbon modified with polyanilinenanofibers and MnO_2 for the separation of the same phenolic compounds. Penget al.,2006 evaluated the presence of surfactant on the eletrochemical behavior of a mixture of catechol and hydroquinone. According to the authors, it was observed that the surfactant is able to shift the anodic peak potential towards lower positive values, thus increasing the analytical signal and providing to a higher sensitivity. In another study, Xuet al., 2011, developed an carbon paste eletrode modified with SBA-15 with β-ciclodextrine. This electrode was characterized and used to simultaneous determination of o-nitrophenol, m-nitrophenol and p-nitrophenol.

Regarding the hydroquinone determination, a survey of literature demonstrates that some electrochemical methods have been developed. Zhao et al., 2007 developed a voltammetric method based on pulse diferential voltammetry using a boron dopped diamond electrode. Hydroquinone was simultaneously determined in the presence of 4-nitrophenol and phenol. Clausen et al., 2013 evaluated the electrochemical performance of carbon nanotubes paste electrode modified with hemin for the hydroquinone determination in pharmaceutical formulations. The use of carbon paste modified with ionic liquid has been proposed for the hydroquinone determination (Zhang et al., 2007). From the results, it was observed that ionic liquid increased significantly the sensitivity of sensor as well as the transfer electron rate and the reversibility of redox process. Electrochemical biosensor has also been employed for hydroquinone determination. Raghu et al., 2013, has shown the use of peroxidase enzyme immobilized on carbon paste chemically modified with poly(L-arginine) and containing silver nanoparticles encapsulated into a sol-gel matrix. Water samples were analysed by proposed method with satisfactory recovery values. The analytical curve of sensor was ranged from 1-150 $\mu mol\ L^{-1}$ and limit of quantification was found to be 1.92 $\mu mol\ L^{-1}$. More recently, Zheng et al.,2013 a sensor based on poly(dopamine)/graphene oxide nanocomposite has been developed and applied to the hydroquinone determination in water samples. A linear range of 1-250 $\mu mol\ L^{-1}$ and limit of detection of 0.62 $\mu mol\ L^{-1}$ were achieved. No interference from Na^{+}, K^{+}, NO^{3-}, SO_4^{2-}, glucose, ascorbic acid, uric acid, glycine, L-cysteine and phenol was observed on the analytical signal of hydroquinone.

Catechol is an isomer of hydroquinone and the analytical methods mostly used for its determination are the spectrophotometric and the chromatographic ones (Lupetti et al., 2004, Liu et al., 2004); however, the electrochemical methods have also been employed for this task. Tarley et al., 2005, synthesized a catechol -imprinted polymer and evaluated its adsorptive properties towards catechol in the presence of structurally analogous molecules 4-chloro -3-methylphenol, 4-aminophenol, 2-cresol, 2-methoxyphenol and 4-chloro -2-methoxyphenol). The proposed method was used to catechol determination in river water and effluent from paper industry. In the study carried out by Sun et al., 2013, differential pulse voltammetry was used to catechol determination using a carbon paste electrode containing ionic liquid as binder and graphene functionalized with poly(cristal violet).

Guaicol is a phenolic compound derived from the degradation of some flavoring, such as eugenol and vanillin, while in the medicaments industry, this compound is used as expectorant. In food samples submitted to the smoking preservation, guaiacol is one of the most responsible molecules for the flavor (Enriquez et al., 2009). In similar way to catechol, the methods commonly used to guaiacol determination are based on chromatography usually coupled to mass spectrometry (Ezquerro et al., 2006, Witthuhn et al., 2012) and some electrochemical approaches can also be reported in the literature. An amperometric biosensor based on immobilization of laccase enzyme on nickel nanoparticles loaded onto oxidized carbon nanotubes has been proposed by Chawla et al., 2012 and used to guaiacol determination in fruit juice. Very low levels at nmol L^{-1} of guaiacol have been achieved through biosensor preparation by using the combination of two enzymes, laccase and tyrosinase (Freire et al., 2002a). In another study, the same authors used the laccase enzyme to development an flow injection system on-line coupled to amperometric determination of guaiacol (Freire et al., 2002b).

4-Nitrophenol is a precursor of pharmaceuticals and agricultural defensives (Zhang et al., 2013). This compound is a sub-product of methyl-parathion degradation, an organophosphorous pesticide widely used (Galeano-Díaz, et al., 2000). In the bloodstream 4-nitrophenol interacts with hemoglobin forming metahemoglobin and may cause cyanosis, confusion and loss of consciousness (U.S. Public Health Service, 2014). Owing to its high toxicity and persistence in the environment, 4-nitrophenol is classified as a dangerous substance, according to US-EPA.

The electrochemical determination of 4-nitrophenol in water samples has been recently demonstrated by using a glassy carbon electrode modified with graphene oxide (Li et al., 2012). Hu et al., 2012, developed an interesting plataform for 4-nitrophenol determination through deposition of molecularly imprinted polymer, zinc oxide nanoparticles, carbon nanotubes and chitosan onto ITO (*indium tin oxide*) substract. A linear range from 0.01-200 μmol L^{-1} and a limit of detection of 0.001 μmol L^{-1} were obtained. The selectivity of sensor towards 4-nitrophenol was assured by the molecularly imprinted polymer due to its electrocatalytic properties.

According to aforementioned, the determination of phenolic compounds in water samples is of paramount importance. In this sence, it is highly desirable the development of analytical methods with low acquisition and maintenance cost, sensitive, selective and capable to simultaneous determination of phenolic compounds. In this context, the use of surfactant in simultaneous electrochemical determination has attracted attention due to simplicity, low cost beyond to corroborates to the actual concept of green chemistry. The surfactants may change the redox potential, the charge transfer and the diffusion coefficients of the electrode process. Moreover, these molecules may avoid the fouling of electrode (Duarte at. al. 2012). Despite this features, so far, no method available for the simultaneous determination of hydroquine, catechol, guaiacol and 4-nitrophenol in water samples has been described in the literature. Therefore, in the present communication we have demonstrated the use of a carbon nanotubes paste sensor in association with measurements performed in presence of a cationic surfactant (cetylpyridiniumbromide, CPB) for the simultaneous determination of hydroquine, catechol, guaiacol and 4-nitrophenol in water samples.

2. Experimental

2.1. Apparatus

The electrochemical experiments were conducted in a three-electrode single-compartment glass cell, including a carbon nanotubes paste sensor as working electrode, Pt plate as auxiliary electrode, and an Ag/AgCl (3.0 mol L^{-1}KCl) as reference electrode. The voltammetric measurements were carried out using a PalmSenspotentiostat/galvanostat controlled with the PalmSens PC software. For comparation studies, a glassy carbon electrode GCE (A = 0.071 cm^2) was used. Before use, it was polished with Al_2O_3 (0.5μm), followed by rinsing with deionized water and then sonicated using alcohol and deionized water mixture solution for 2 min.

2.2. Reagents

Hydroquine, cathecol, guaiacol, 4-nitrophenol and cetylpyridinium bromide were obtained from Sigma-Aldrich. Multi-walled carbon nanotubes (MWCNTs; of 10–40 nm in diameter and 5-20 μm in length; purity: 93%) was obtained from CNT Co. Ltd., Korea. The phosphate buffer solution was prepared from its respective salt (NaH_2PO_4 99-102%) from Merck. Mineral oil (Nujol®) was acquired in local drug stores. Britton-Robison (BR) buffer solution was prepared by mixing 0.04 mol L^{-1} in acetic, orthophosphoric, and boric acids, with pH adjusted to 5.0 with a 2.0 mol L^{-1}NaOH solution. All chemicals were analytical grade, and the solutions were prepared using ultra-purified water (resistivity > 18 MΩ cm) supplied by a Milli-Q system (Millipore®).

2.3. Preparation of Carbon Nanotube Paste Sensor

The carbon nanotubes paste sensor was preparedby mixing 10.0 mg of multiwall carbon nanotubes (MWCNTs) with 23.3 mg of mineral oil ina Petri dishwith a steel spatulafor 15 minutes until a homogeneous paste was obtained. The paste at a ratio of 30:70% (w/w) (MWCNT/Nujol) was than carefully forced into a cavity (3 mm diameter; 1 mm depth) at the end of a Teflon® tube. The electrical connection was provided by a disc of graphite connected to the paste in the inner hole of the tube. The surface of the paste electrode was smoothed and rinsed carefully with Milli-Q water. The paste was activated by cyclic voltammetry in the potential ranging from -1.0 up to 1.0 V with scan rate of 75 mV s^{-1} during 30 minutos.

2.4. Experimental Procedure

Cyclic and linear voltammetric measurements were carried out at a electrochemical cell of 15 mL capacity without removing oxygen under pH 7.0 (0.1 mol L^{-1} phosphate buffer) containing the working electrode (carbon nanotubes paste electrode), Ag/AgCl (3.0 mol L^{-1}) reference electrode and platinum as counter electrode. The cationic surfactant cetylpyridinium bromide (CPB) concentration optimized was 0.45 mmol L^{-1}. The experimental parameters of cyclic voltammetry were the following: step potential of 5 mV, scan rate of 30 mV s^{-1} while a scan rate of 120 mV s^{-1} was used to linear voltammetry.

3. Results and Discussion

3.1. Influence of CPB on the Oxidation of Phenolic Compounds

Figure 2 shows the influence of the surfactant on the oxidation process of hydroquine, cathecol, guaiacol and 4-nitrophenol. As observed, in the presence of 0.45 mmol L^{-1} CBP the anodic peak current of phenolic compounds increases significantly and the peak potential shifts toward less positive potential. Such result demonstrates that the electrochemical oxidation of phenolic compounds under cationic surfactant medium is electrocalyzed. It is very well known that surfactants in solution may promote the aggregation and adsorption on the electrode surface at a film form.

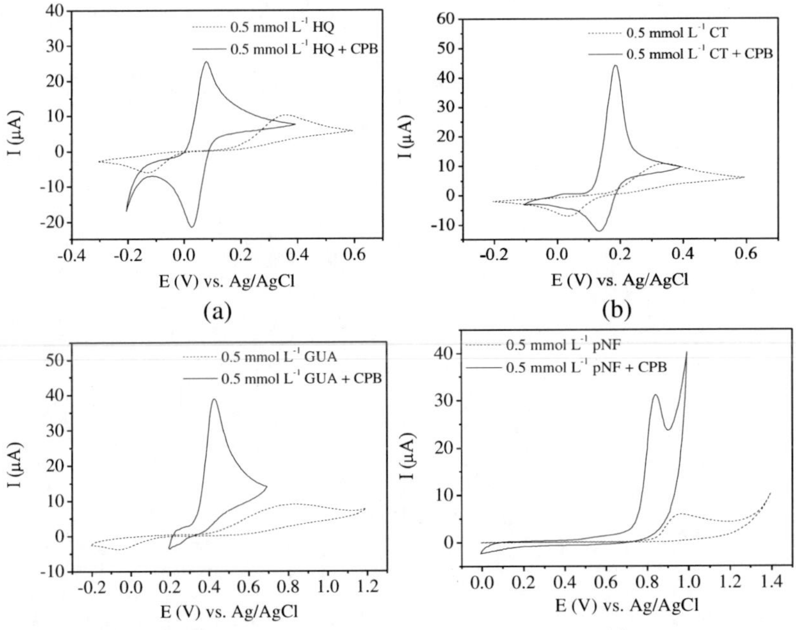

Figure 2. Cyclic voltammograms of (a) hydroquinone, (b) catechol, (c) guaiacol and (d) 4-nitrophenol recorded on carbon nanotubes paste sensor in the absence and presence of 0.45 mmol CPB. Conditions: phenolic compounds at 500 μmol L^{-1} concentration, support electrolyte – phosphate buffer at pH 7.0.

Therefore, the formation of positively charged hydrophilic film organized towards the water bulk phase at pH 7.0 and the partial ionization of phenolic compounds under neutral pH, makes possible the electrostatic interaction of phenolic compounds with surface electrode, which justify the electrocatalysis decreasing the overpotential and increasing the electron transfer rate (Dos Reis, 2008).

3.2. Comparative Study of Carbon Nanotubes Paste Sensor with Glassy Carbon Electrode

In order to emphasize the benefits of carbon nanotubes as a nanostructured material, experiments using glassy carbon electrode also in the presence of cationic surfactant were conducted. The results are showed on Figure 3.

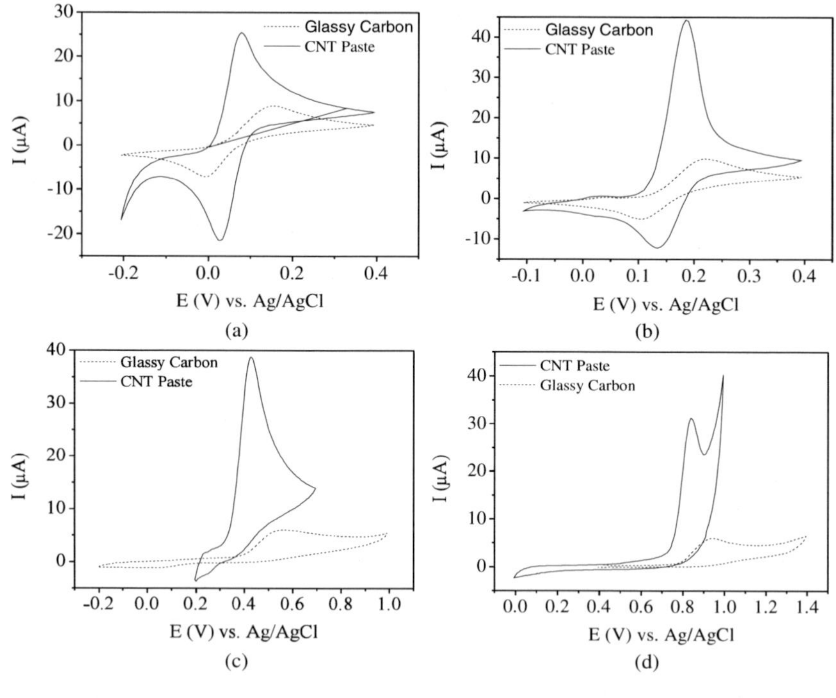

Figure 3. Cyclic voltammograms of (a) hydroquinone, (b) catechol, (c) guaiacol and (d) 4-nitrophenol recorded on carbon nanotubes (CNT) paste sensor and glassy carbon electrode in the presence of 0.45 mmol CPB. Conditions: phenolic compounds at 500 μmol L^{-1}concentration, support electrolyte – phosphate buffer at pH 7.0.

As can be seen, the use of carbon nanotubes paste sensor in comparison to glassy carbon electrodes promotes considerable increase on the peak currents and decrease in the overpotential, thus emphasizing the great advantages of MWCNT in detriment of glassy carbon electrode. Apart from the electrical properties of carbon nanotubes, which justify their better analytical response in comparison to glassy carbon electrode, these materials also possess high surface area, thus favoring the adsorption of cationic surfactant.

3.3. Simultaneous Determination of Hydroquinone, Catechol, Guaiacol and 4-Nitrophenol

The feasibility of the carbon nanotubes paste sensor for the simultaneous determination of hydroquinone, catechol, guaiacol and 4-nitrophenol in the presence of cationic surfactant was checked and compared with glassy carbon electrode.

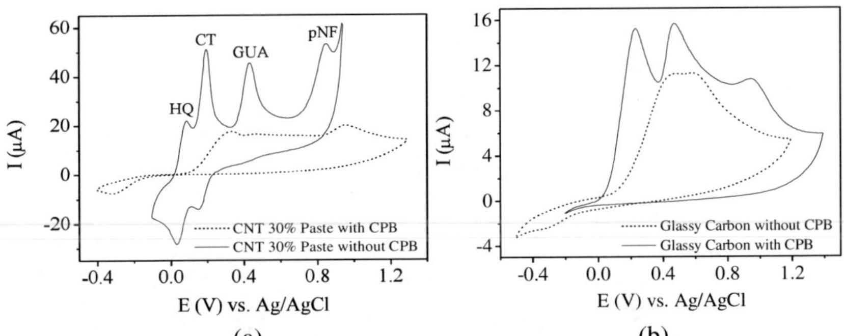

Figure 4. Cyclic voltammograms of hydroquinone, catechol, guaiacol and 4-nitrophenol recorded on carbon nanotubes paste sensor in the presence and absence of 0.45 mmol CPB (a) and glassy carbon electrode in the presence and absence of 0.45 mmolCPB. Conditions: phenolic compounds at 500 µmol L^{-1} concentration, support electrolyte – phosphate buffer at pH 7.0.

As one can see from Figure 4, the use of carbon nanotubes paste sensor in the presence of CPB makes possible the simultaneous determination of phenolic compounds with a peak potential separation (ΔE_{ap}) at least 112 mV, suggesting the selective measurement of one compound in the presence of other specie or the feasible simultaneous determination of these four compounds at the sensor. When the glassy carbon electrode was used even in the presence of CPB, it was observed overlapping oxidation peaks of hydroquinone and catechol and a great broadening of peak potential of 4-nitrophenol, thus making impossible the reliable simultaneous determination of four compouns using this commercial available electrode. Moreover, no resolved peak potentials were observed for phenolic compounds using both carbon nanotubes paste sensor and glassy carbon electrode in the absence of CPB, justifying again the effect of surfactant on the peak separation and increasing the analytical currents.

3.4. Influence of Carbon Nanotube and Mineral Oil

In order to determine the most favorable conditions for preparing the carbon nanotube paste electrode, the influence of the amount of carbon nanotube and mineral oil were investigated on the electrochemical response of sensor for simultaneous determination of phenolic compounds.

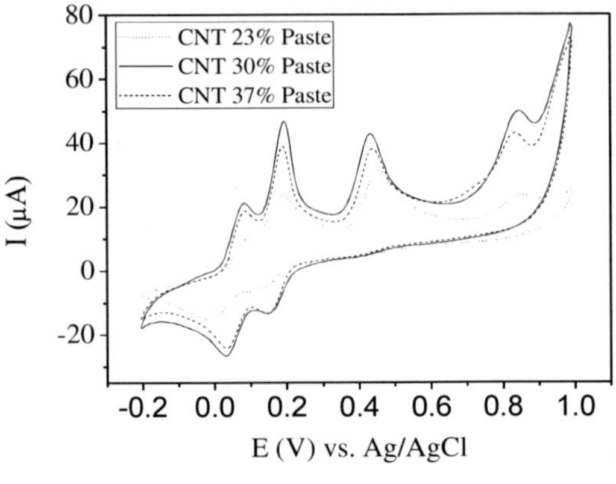

Figure 5. Comparison of cyclic voltammograms for phenolic compounds at three different working electrodes built with MWCNT and mineral oil. Concentration of phenolic compounds: 500 μmol L^{-1}. Measurements carried out in 0.1 mol L^{-1} phosphate buffer solution. Potential scan rate: 30 mV s^{-1}.

Different ratios MWCNT/Mineral Oil (23%/77%, 30%/70%, 37%/63% m/m) were investigated and the cyclic voltammograms are depicted on Figure 5. As observed, the use lower amount of MWCNT in the composition of the carbon paste decreases the anodic peak current of phenolic compounds as well as the peak separation, probably due to decreases in the conductivity of the sensor.

A similar electrochemical behavior was observed by using the amount of 37% and 30% (m/m) of MWCNT, but we have chosen as the best condition the amount of 30% (m/m). This composition is very easy to handle; the paste is physically stable into the cavity of sensor; is highly homogeneous and promotes a smooth surface of sensor.

3.5. Influence of CPB Surfactant on Simultaneous Determination of Hydroquinone, Catechol, Guaiacol and 4-Nitrophenol

The influence of CPB concentration on simultaneous determination of phenolic compounds was investigated from 0.15 up to 0.90 mmol L^{-1} (Figure 6).

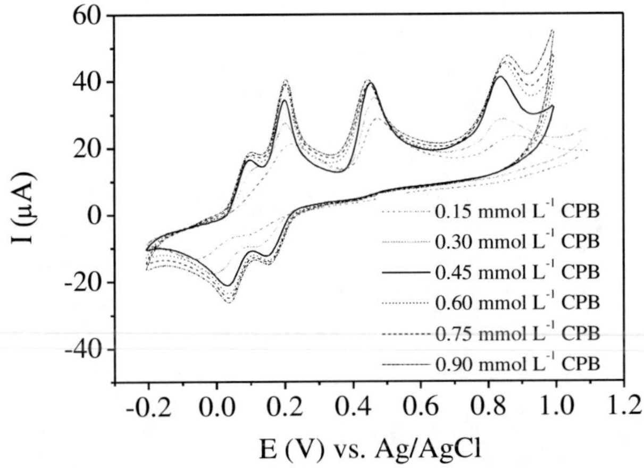

Figure 6. Cyclic voltammograms recorded for the simultaneous determination of hydroquinone, catechol, guaiacol and 4-nitrophenol in the presence of increasing CPB concentration. Concentration of phenolic compounds: 500 μmol L^{-1}. Measurements carried out in 0.1 mol L^{-1} phosphate buffer solution. Potential scan rate: 30 mV s^{-1}.

According to results, it can be observed that using lower CPB concentrations (0.15 and 0.30 mmol L^{-1}) the anodic peak potentials of phenolic compounds are shifted towards more positive values, the peak currents are decreased and there exist an overlapping of peak potential of catechol and hydroquinone. On the other hand, for higher concentration varying from 0.45 up to 0.90 mmol L^{-1} the analytical response of sensor towards phenolic compounds leveled off. Thus, the CPB concentration of 0.45 mmol L^{-1} was chosen as optimum condition.

3.6. Influence of the pH on the Oxidation of Hydroquinone, Catechol, Guaiacol and 4-Nitrophenol

The influence of pH on anodic peak potential of phenolic compounds was examined under the range varying from 2 up to 8 (Figure 7). For these assays, cyclic voltammetry was used and the records were performed in phosphate buffer solution at 0.1 mol L^{-1} concentration.

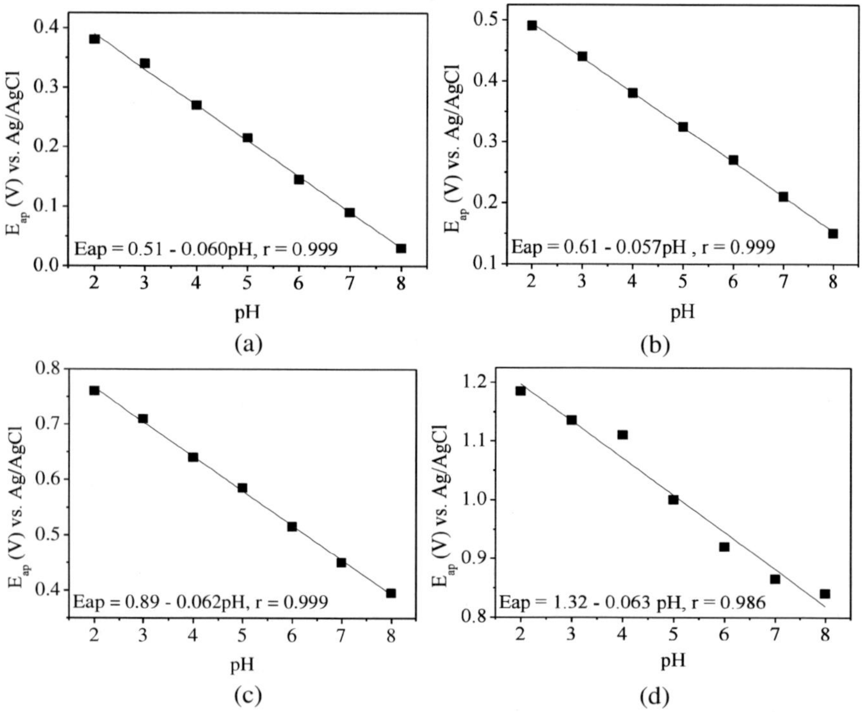

Figure 7. Influence of the pH on the anodic peak potential. Concentration of phenolic compounds: 500 μmol L^{-1}. Measurements carried out in 0.45 mmol L^{-1} CPB. Potential scan rate: 30 mV s^{-1}.

A linear correlation was observed under pH range studied for anodic peak potential *vs* pH with slopes of 0.060, 0.057, 0.062 and 0.063 V/pH for respective compounds hydroquinone, catechol, guaiacol and 4-nitrophenol. As these values are very close to 0.059 V/pH at 25 °C, which correspond to a Nernstian system (Papouchado et al., 1975), we can infer that the number of protons and electrons involved in the electrode reaction of phenolic compounds is the same.

This study also showed that peak currents of phenolic compounds are pH dependent, whose the best results were achieved for pH 7.0, except for 4-nitrophenol, which gives the best response at pH 4.0. Based on these findings, a study for obtaining the optimum pH for the simultaneous determination of phenolic compounds was carried out.

Figure 8 shows the pH dependence on the separation of anodic peak potential. Under acid medium (pH 5.0 and 6.0) no satisfactory peak separation was observed, especially for hydroquinone and catechol, with overlapping signal, while measurements at pH 7.0 and 8.0 provides a good peak separation. More alkaline medium (pH 9.0) (data not shown) was also investigated, but a significant decrease on potentials and peak currents was observed.

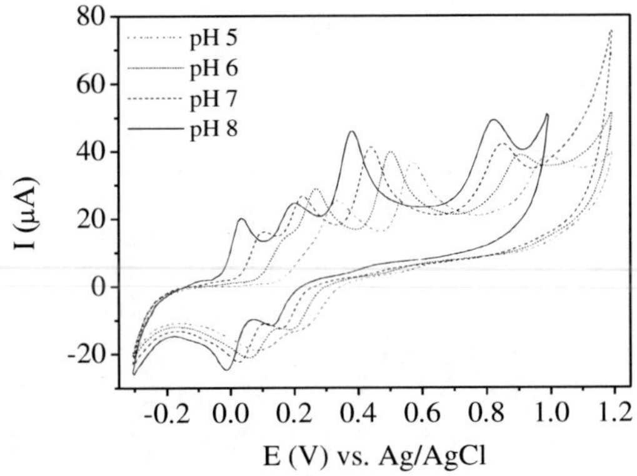

Figure 8. Voltammetric profile of simultaneous determination of hydroquinone, catechol, guaiacol and 4-nitrophenol under different pH values.

3.7. Effect of Scan Rate Potential on the Oxidation of Hydroquinone, Catechol, Guaiacol and 4-Nitrophenol

Cyclic voltammograms were recorded over a range of scan rate varying from 5 up to 100 mV s^{-1}. In order to check whether the electron transfer process is diffusion or adsorption controlled, plots logarithm of the anodic peak current [log(Ipa)] versus the logarithm of the scan rate [log(ν)] for the phenolic compounds were recorded. These experiments were carried out in phosphate buffer solution (pH 8.0) containing 0.45 mmol L^{-1} CPB. When the slope of linear plot is close to 0.5 the electron transfer process is diffusion controlled, while a slope close to 1.0 the adsorption is the limiting step of process. Hence, for hydroquinone and guaicol the slopes were found to be 0.74 and 0.46, respectively, thus suggesting a diffusion-controlled mechanism. As regards the catechol and 4-nitrophenol the obtained values were 0.88 and 0.96, respectively, thus for these phenolic compounds the electron transfer process is adsorption controlled (Gosser, 1994). The results herein found corroborates with the calculation of diffusion coefficients experimentally determined as will be further demonstrated. The diffusion coefficient of each phenolic compound was determined from chronoamperometric studies. The chronoamperometric measurements were recorded by setting the electrode potential at 0.09, 0.210, 0.450, 0.860 V for hydroquinone, catechol, guaiacol and 4-nitrophenol, respectively and varying the concentration of compounds from 0.05 up to 0.30 mmol L^{-1}. Using the Cottrell equation (1) (Bard & Faulkner, 2001), linear plots of I_{ap}vs $t^{-1/2}$ were recorded (Figure 9).

$$I_{ap} = \frac{nFAC_o D_o^{1/2}}{\pi^{1/2} t^{1/2}} \quad (1)$$

where, n = 2 F is the Faraday constant (96500 C), A is active area of electrode (cm^2), D_o is the diffusion coefficient ($cm^2\ s^{-1}$) and C_o is the concentration (mol cm^{-3}). From the average of slopes, and using the Cottrell equation, the average value of the diffusion coefficients were determined as being 5.45×10^{-6}, 3.35×10^{-5}, 7.35×10^{-5} and $8.05 \times 10^{-5} cm^2\ s^{-1}$ for hydroquinone, catechol, guaiacol and 4-nitrophenol, respectively. It worth remember that before direct use of Cottrell equation, the active area of electrode ($4.66 \times 10^{-2}\ cm^2$) was determined from the slope of plot Ipvs$v^{1/2}$, according to Randles-Sevcik equation (2) (Bard & Faulkner, 2001) using 5.0 mmol L^{-1} $K_4Fe(CN)_6$ in 1.0 mol L^{-1}KCl electrolyte.

$$I_{pa}=2.69 \times 10^5 n^{3/2} A D^{1/2} v^{1/2} C \quad (2)$$

where, n =1 and D is the diffusion coefficient of [1] $K_4Fe(CN)_6$ of $7.6 \times 10^{-6}\ cm^2\ s^{-1}$. Moreover, the number of electrons involved in the electrochemical oxidation of phenolic compounds were determined by applying the equation

$E_p - E_{p/2} = 47.7$ mV/αn for totally irreversible system (4-nitrophenol) and $E_p - E_{p/2} = 59$mV/n for reversible (hydroquinone) and quasi-reversible system (catechol and guaiacol) (Greefet al., 1990). The n value was calculated as 1.63, 1.73, 1.55 and 1.77 for hydroquinone, catechol, guaiacol and 4-nitrophenol, respectively, thus indicating that in fact the number of electron transferred in the rate-determining step should be 2.0.

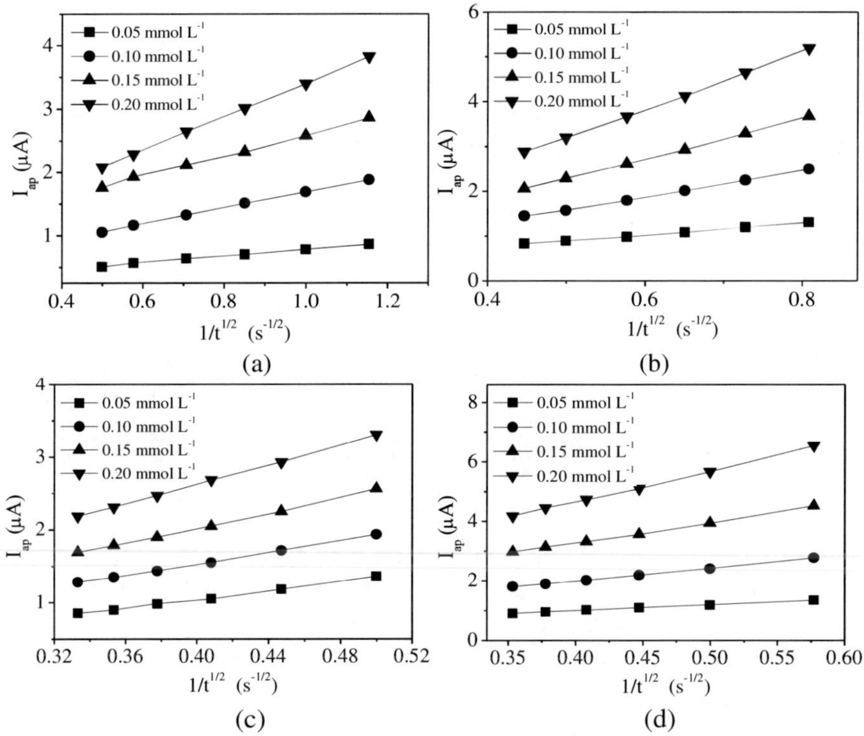

Figure 9. Linear plots I_{ap}vs $t^{1/2}$ from Cottrell's equation. (a) hydroquinone, (b) catechol (c) guaiacol and (d) 4-nitrophenol.

3.8. Multivariate Optimization of Simultaneous Determination of Hydroquinone, Catechol, Guaiacol and 4-Nitrophenol

The best condition for the simultaneous deteterminaton of hydroquinone, catechol, guaiacol and 4-nitrophenol was achieved by using a 2^4 full factorial design associated with multi-response. For this task, cyclic voltammetry was used as electrochemical technique under a potential range varying from -0.2 up to 1.0 V vs Ag/AgCl and the concentration of phenolic compounds and CPB were fixed at 0.3 mmol L^{-1} and 0.45 mmol L^{-1}, respectively. Table 1 shows the factors studied and their respectivelevels and the results of this factorial design are shown in Table 2. The assays were performed in duplicate and the peak current of phenolic compounds was taken as analytical response. However, in order to simultaneously improve the four responses *i.e.* analytical signals of hydroquinone, catechol, guaiacol and 4-nitrophenol, the peak currents were converted into a desirability function and after in overall desirability. Our research group has previously demonstrated the usefulness of the desirability function for the optimization of multi-response assays in proposed voltammetric method (Duarteet al., 2012, Hudari et al. 2013).

Table 1. Factors and their levels used in the 2^4 factorial design

Factors	High level (+)	Low level (-)
1: pH	8	7
2: ET (electrolyte type)	BR buffer	Phosphate buffer
3: EC (electrolyte concentration)	0.1 mol L^{-1}	0.01 mol L^{-1}
4: Surfactant type (ST)	CPB	CTAB

BR – Britton-Robinson; cetylpyridinium bromide; CTAB – *cetyltrimethylammonium bromide.*

Firstly, the individual desirability (d_i) for each phenolic compound was expressed as a dimensionless value ranging from 0 (undesired response) to 1 (desired response) using the equation (3).

$$d_i = \left(\frac{y - L}{H - L} \right), \tag{3}$$

wherey is the I_{pa} value obtained experimentally, L and H are the lowest and the highest values, respectively, among those obtained in all experiments. After determining the individual desirability, the overall desirability was calculated according to equation (4).

$$O.D = \sqrt[m]{d_1 d_2 ... d_m} \tag{4}$$

where*m*isthe number of responses studied for each experimentduring the optimization process, in this case *m* was changed from 3 to 4 responses.

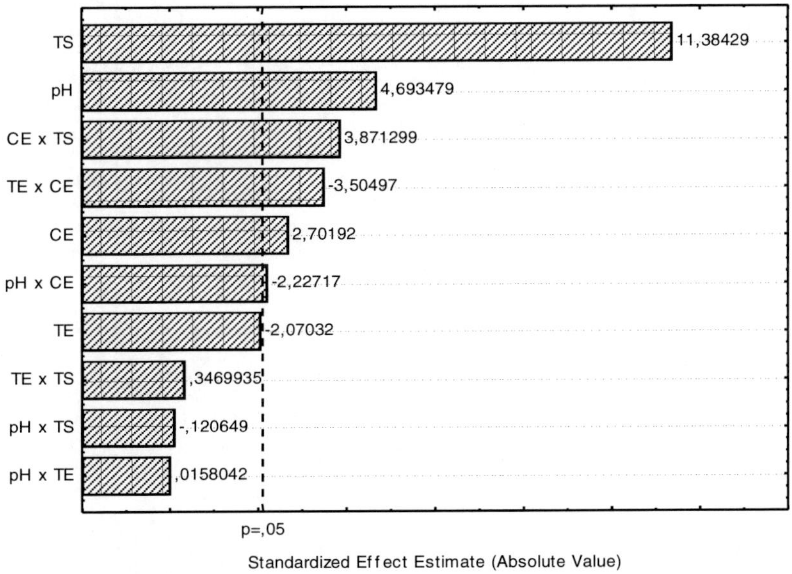

Figure 10. Pareto chart obtained from the 2^4 full factorial design.

The results obtained from the 2^4 factorial design using overall desirability as multi-reponse were evaluated by analysis of variance (ANOVA) and graphically represented by a Pareto chart (Figure 10). In this chart, when the bar exceeds the dotted vertical reference line, which defines the 95 % confidence interval, the factor or interaction is considered significant (Diniz et al., 2013). As observed from the Pareto chart, surfactant type plays an important role on the analytical response with positive effect (11.38), thus indicating that CPB surfactant provides a better performance for the simultaneous determination of phenolic compounds in comparison with CTAB.

A similar effect was observed for pH and electrolyte concentration *i.e.* the better response were obtained at higher levels (pH 8.0 and concentration of 0.1 mol L^{-1}).

It was not observed influence of electrolyte type (BR and phosphate) on the analytical currents; however a better peak separation was verified using BR.

Table 2. Factorial design and analytical responses obtained (current, desirability and global desirability)

Assays	pH	ET	EC	ST	Corrent / μA				Desirability (d_i)				Global desirability (D_g)
					HQ	CT	GUA	pNF	HQ	CT	GUA	pNF	
1	-	-	-	-	*	5.774	4.672	2.341	*	0.530	0.082	0.261	0.221
2	+	-	-	-	*	5.653	5.070	1.872	*	0.512	0.148	0.161	0.228
3	-	+	-	-	*	6.106	5.570	1.725	*	0.580	0.231	0.130	0.258
4	+	+	-	-	*	5.049	6.576	2.027	*	0.421	0.397	0.194	0.318
5	-	-	+	-	*	4.449	4.383	1.268	*	0.330	0.034	0.032	0.057
6	+	-	+	-	*	8.435	8.248	2.586	*	0.933	0.675	0.312	0.581
7	-	+	+	-	*	2.733	6.492	2.442	*	0.070	0.383	0.282	0.196
8	+	+	+	-	*	2.355	7.948	2.509	*	0.013	0.625	0.297	0.077
9	-	-	-	+	3.005	8.060	9.277	2.467	0.163	0.876	0.845	0.288	0.427
10	+	-	-	+	3.133	6.812	8.073	3.277	0.204	0.687	0.646	0.460	0.448
11	-	+	-	+	2.572	8.370	8.188	2.496	0.022	0.923	0.665	0.294	0.148
12	+	+	-	+	4.166	7.571	8.631	5.524	0.541	0.802	0.738	0.939	0.739
13	-	-	+	+	3.552	8.856	10.028	4.418	0.341	0.997	0.970	0.704	0.693
14	+	-	+	+	3.390	7.551	9.183	4.334	0.288	0.799	0.830	0.686	0.601
15	-	+	+	+	3.845	5.430	9.794	4.077	0.436	0.478	0.931	0.631	0.591
16	+	+	+	+	5.456	3.303	9.025	3.968	0.961	0.156	0.803	0.608	0.520

* Analytical current not recorded; TS – Surfactant type; EC – Electrolyte concentration; ET – Electrolyte type; HQ – Hydroquinone; CT – Catechol; GUA – Guaiacol; pNF – 4-nitrophenol.

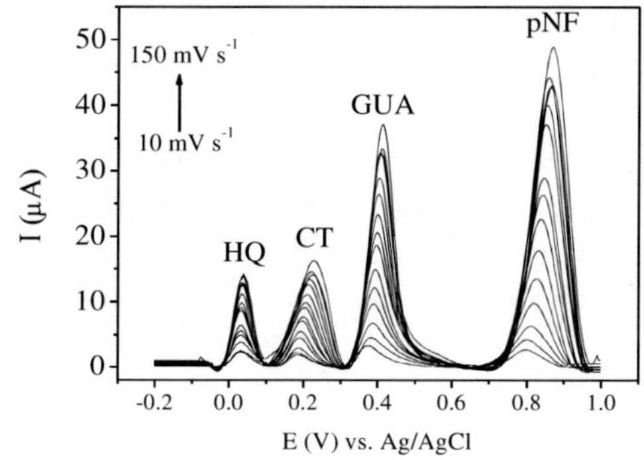

Figure 11. Linear sweep voltammograms recorded under different scan rate potential.

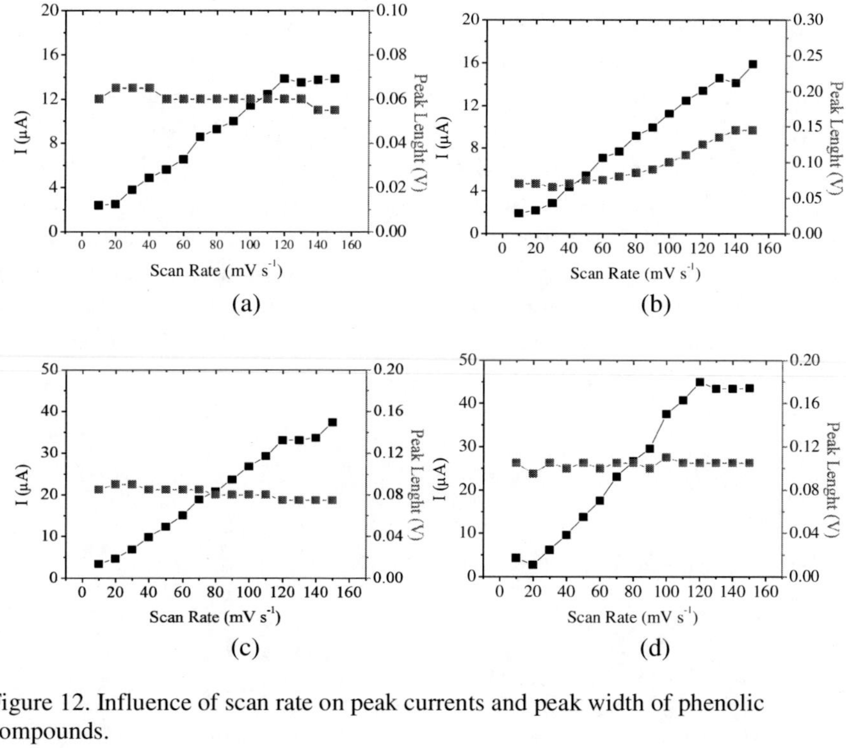

Figure 12. Influence of scan rate on peak currents and peak width of phenolic compounds.

Thus, the optimized condition for the successful separation of four phenolic compounds was: BR as electrolyte support, pH 8.0, CPB as surfactant and electrolyte concentration at 0.1 mol L^{-1}.

After final optimization, linear sweep voltammetry (LSV) was employed as analytical technique for the simultaneous determination of phenolic compounds. In this sence, the influence of scan rate potential on the performance of LSV was evaluated in the range 10-150 mV s^{-1} (Figure 11). It was verified that with increasing the scan rate, the peak currents of phenolic compounds increases until a plateau with 120 mV s^{-1} (Figure 12). From this figure, it can also be observed that peak width under the wide scan rate evaluated is practically not affected. Thus, the scan rate of 120 mV s^{-1} for LSV was adopted in this study.

3.9. Analytical Performance of LSV Method for the Simultaneous Determination of Hydroquinone, Catechol, Guaiacol and 4-Nitrophenol

Figure 13 depicts the LSV records for phenolic compounds under increasing concentrations yielding the following equations i(μA) = 0.999 + 0.068[HQ] (r = 0.998), i(μA) = 0.255 + 0.066[CT], (r = 0.997), i(μA) = 0.271 + 0.157[GUA], (r = 0.999), i(μA) = 0.850 + 0.160[pNF],(r = 0.997). The limits of detection (LOD) and quantification (LOQ) were calculated according to IUPAC recommendation based on the slope of the calibration curves using the following equations: *LOD=3std/M* and *LOQ = 10std/M*, where *std* is the standard deviation of 10 blank determinations, and *M* is the slope of the calibration curve (Long &Winefordner, 1983).

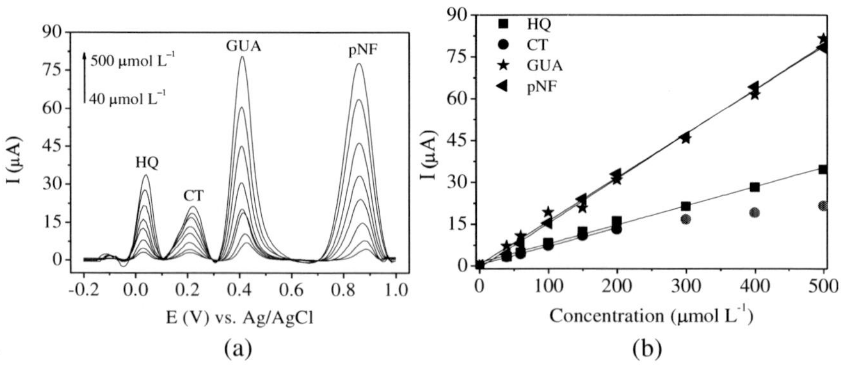

Figure 13. (a) LSV for the simultaneous determination of phenolic compounds; (b) analytical curves for phenolic compounds.

Table 3. Results obtained for the precision and accuracy of the proposed method

Parameter	Days	Concentration (μmol L^{-1})							
		HQ	CT	GUA	pNF	HQ	CT	GUA	pNF
Nominal concentration		60.00	60.00	60.00	60.00	400.00	150.00	400.00	400.00
Analyzed concentration (n=10)	1	59.28	60.60	61.74	59.21	406.26	152.14	392.14	408.05
	2	60.19	60.68	61.77	58.83	401.47	150.71	394.58	405.74
Precision (relative standard deviation, %)	1	1.53	1.94	1.99	2.53	2.93	2.02	3.50	3.32
	2	2.02	2.04	1.21	1.30	2.70	2.48	2.42	2.99
Accuracy (relative error, %)	1	-1.20	0.99	2.90	-1.32	1.57	1.43	-1.97	2.01
	2	0.32	1.14	2.94	-1.95	0.34	0.47	-1.36	1.43

HQ= hydroquinone; CT = catechol; GUA = guaiacol; pNF = 4-nitrophenol.

Therefore, the LOD were found to be 3.43, 1.72, 1.10, and 4.96 μmol L^{-1} for hydroquinone, catechol, guaiacol and 4-nitrophenol, respectively.

For the LOQ, the values were found to be 11.43, 5.73, 3.67 and 16.53 μmol L^{-1}. The obtained results for within-day and between-day precision and accuracy at concentration levels of 60 and 400 μmol L^{-1} for hydroquinone, guaiacol and 4-nitrophenol and 60 and 150 μmol L^{-1} for catechol are presented in Table 3. As observed, the proposed method is precise and accurate within the desired range.

3.10. Application of Proposed Method for the Simultaneous Determination of Hydroquinone, Catechol, Guaiacol and 4-Nitrophenol in Water Samples

Tap water and mineral were obtained at State University of Londrina and local supermarked, respectively. After collection, the samples had the pH adjusted to 8.0 with BR buffer solution and immediately analyzed by proposed method. Lake water samples were collected in polypropylene bottles from Igapó Lake, Londrina, Brazil. Afterwards, the samples were filtered under vacuum using 0.45 μm cellulose acetate membranes and the pH was adjusted to 8.0 with BR buffer solution. The feaseability of method for the simultaneous determination of hydroquinone, catechol, guaiacol and 4-nitrophenol in these water samples was checked from addition and recovery tests after adding known phenolic compounds amounts to the samples. As observed from Table 4, good results were obtained (>92%), thus confirming the interference-free applicability of method.

Table 4. Determination of phenolic compounds in water samples

Samples	Analyte	Phenolic compounds added (μmol L^{-1})	Phenolic compounds found (μmol L^{-1})[a]	Recovery (%)
Lake water	HQ	60.0	57.6±1.1	96
		120.0	126.9±1.5	106
	CT	60.0	60.4±1.5	101
		120.0	118.2±1.0	99
	GUA	60.0	55.4±1.8	92
		120.0	116.3±2.0	97
	pNF	60.0	64.0±1.2	107
		120.0	121.8±2.0	102
Tap water	HQ	60.0	63.2±1.1	105
		120.0	112.8±1.2	94
	CT	60.0	61.7±1.7	103
		120.0	118.3±1.3	99
	GUA	60.0	62.9±1.9	105
		120.0	112.5±1.3	93
	pNF	60.0	64.5±1.7	108
		120.0	114.1±1.4	95
Mineral water	HQ	60.0	61.7±2.0	103
		120.0	122.6±0.4	102
	CT	60.0	60.53±1.8	101
		120.0	117.1±2.2	98
	GUA	60.0	59.7±1.1	99
		120.0	125.5±2.0	105
	pNF	60.0	62.7±1.2	105
		120.0	129.2±1.4	108

[a]Results are expressed as mean value ± standard deviation based on three replicates (n=3).

CONCLUSION

The preparation and application of a carbon nanotube paste sensor was successfully carried out for the simultaneous determination of four phenolic compounds (hydroquinone, catechol, guaiacol and 4-nitrophenol) in the presence of cationic surfactant. It was demonstrated that measures in surfactant medium significantly increases the anodic peak current of phenolic compounds as well as provides a better peak separation. The most favorable feature of LSV method is its ease to be performed, presents satisfactory limits of detection, low cost to be implemented, good precision, is able to determine the four phenolic compounds in water samples even those environmental samples (lake water) and employs the green chemistry concept because makes use of surfactant in the measures.

REFERENCES

Afkhami. A.; Ghaedi. H.; Madrakian. T.; Nematollahi. D.; Mokhtari. B.; *Talanta* 2014. *121*. 1.

Barcelo. D.; *Trends Anal. Chem.*2003. *22*. 15.

Bard. A. J.; Faulkner. L. R. *Electrochemical Methods. Fundamentals and Applications*. Second ed.. Wiley. New York 2001.

Bensaid. F. F.; Wietzerbin. K.; Martin. G. J.; *J. Agric. Food Chem.* 2002. *50*. 6271.

Bussy. U.; Ferchaud-Roucher. V.; Tea. I.; Krempf. M.; Silvestra. V.; Boujtita. M.; *Electrochim. Acta*. 2012. *69*. 351.

Chawla. S.; Rawal. R.; Sharma. S.; Pundir. C. S.; *Biochem. Eng. J.* 2012. *68*. 76.

Clausen. D. N.; Duarte. E. H.; Pereira. A. C.; Kubota. L. T.; Tarley. C. R. T. Em*Biomimetics Research*; Gotsiridze-Columbus. N.; Rosa S.; 1^{ed}. Nova Iorque: Nova Publishers. 2013. 111.

CONAMA – Concelho Nacional do Meio Ambiente. Resolução nº 430 de 13 de Maio de 2011. Disponível em <http://www.mma.gov.br/port/conama/legiabre.cfm?codlegi=646>. Acesso em 15 Jan. 2014

Diniz. K. M.; Segatelli. M. G.; Tarley. C. R. T. *Reac. Funct. Polym* 2013.*76*. 838.

Dos Reis. A. P.; Tarley. C.R.T.; Kubota. L. T.; *J. Braz. Chem. Soc.* 2008. *19*. 1567.

Duarte. E. H.; Kubota. L. T.; Tarley. C. R. T.; *Electroanalysis* 2012. *24*. 2291.

Eggen. T.; Moeder. M.; Arukwe. A.; *Sci. Total Environ.* 2010. *408*. 5147.

Enriquez. M. E. M.; Villar. A. D.; Chauvet. D.; Valle. A. L.; Pompeyo. M. S.; Sepúlveda. A. E. C.; *Proc. West. Pharmacol. Soc.* 2009. *52*. 92.

Ezquerro. O.; Garrido-López. A.; Tena. M. T.; *J. Chromatogr. A* 2006.*1102*. 18.

Freire. R. S.; Duran. N.; Kubota. L. T.; *Anal. Chim. Acta* 2002. *463*. 229b.

Freire. R. S.; Thongngamdee. S.; Durán. N.; Wang. J.; Kubota. L. T.; *Analyst* 2002. *127*. 258[a].

Galeano-Díaz. T.; Guiberteau-Cabanillas. A.; Mora-Díez. N.; Parrilla-Vázquez. P.; Salinas-López. F.; *J. Agri. Food Chem.* 2000.*48*. 4508.

Gao. W.; Tiju. W. W.; Wei. J.; Liu. T.; *Talanta* 2014. *120*. 484.

Gosser. D. K. *Cyclic Voltammetry*; VCH Publishers: New York. 1994.

Greef. R.; Peter. R.; Peter. L. M.; Pletcher. D.; Robinson. J. *Instrumental Methods in Electrochemistry*. Ellis Horwood Limited. West Sussex 1990.

Hu. Y.-F.; Zhang. Z.-H.; Zhang. H.-B.; Luo. L.-J.; Yao. S.-Z.; *Thin Solid Films* 2012. *520*. 5314.

Hudari. F. F.; Duarte. E. H.; Pereira. A. C.; Dall`Antonia. L.H.; Kubota. L. T.; Tarley. C.R.T. *J. Electroanal. Chem.* 2013. *696. 52.*

Karadas. N.; Ozkan. S. A.; *Talanta* 2014. *119*. 248.

Li. J.; Kuang. D.; Feng. Y.; Zhang. F.; Xu. Z.; Liu. M.; *J. Hazard. Mater.* 2012. *201-202*. 250.

Liu. R.; Zhou. J. L.;Wilding. A.; *J. Chromatogr. A* 2004. *1022*. 179.

Long. G. L.; Winefordner. J. D. *Anal. Chem.* 1983.*55*. 712.

Lupetti. K. O.; Rocha. F. P. R.; Fatibello-Filho. O.; *Talanta*2004.*62*. 463.

Madrakian. T.; Haghshenas. E.; Afkhami. A.; *Sens. Actuators. B* 2014. *193*. 451.

Muñoz. I.; Gómez. M. J.; Molina-Díaz. A.; Huijbregts. M. A. J.; Fernández-Alba. A. R.; García-Calvo. E.; *Chemosphere* 2008. *74*. 37.

Murray. K. E.; Thomas. S. B.; Badour. A. A.; *Environ. Pollut.*2010. *158*. 3462.

Papouchado. L.; Sandford. R. W.; Petrie. G.; Adams. R. N; *J. Electroanal. Chem.* 1975.*65*. 275.

Peng. J.; Gao. Z.-N.; *Anal. Bioanal. Chem.* 2006. *384*. 1525.

Petrovic. M.; Gonzales. S.; Barceló. D.; *Trends Anal. Chem.* 2003. *22*. 685.

Prathap. M. U. A.; Satpati. B.; Srivastava. R.; *Sens. Actuators. B* 2013. *186*. 67.

Raghu. P.; Reddy. T. M.; Reddaiah. K.; Jaidev. L. R.; Narasimha. G.; *EnzimeMicrob. Technol.* 2013. *52*. 377.

Reemtsma. T.; Quintana. J. B.; Rodil. R.; Gárcia-López. M.; Rodríguez. I.; *Trends Anal. Chem.* 2008.*27*. 727.

Rosatto. S. S.; Freire. R. S.; Durán. N.; Kubota. L. T.; *Quim. Nova* 2001. *24*. 77.

Silva. C. G. A.; Collins. C. H.; *Quim. Nova* 2011. *34*. 665.

Sun. W.; Wang. Y.; Lu. Y.; Hu. A.; Shi. F.; Sun. Z.; *Sens. Actuators. B* 2013. *188*. 564.

Tarley. C. R. T.; Kubota. L. T.; *Anal. Chim. Acta* 2005. *548*. 11.

U. S. Public Health Service. Agency fo Toxic Substances and Disease Registry. Toxicological Profile for Nitrophenols. Disponível em: < http://www.atsdr.cdc.gov/toxprofiles/tp50.pdf >. Acesso em 15 Jan. 2014.

Urtiaga. A. M.. Pérez. G.. Ibáñez. R.. Ortiz. I.; *Desalination* 2013.*331*. 26.

Wang. Q.; Arash. B.; *Comput. Mater. Sci.* 2014. *82*. 350.

Wei. L.; Lu. D.; Wang. J.; Wei. H.; Zhao. J.; Geng. H.; Zhang. Y.; *Sens. Actuators. B* 2014. *190*. 529.

Witthuhn. R. C.; Merwe. E. V. D.; Vanter. P.; Cameron. M.; *Int. J. Food Microbiol.* 2012. *157*. 113.

Xu. X.; Liu. Z.; Zhang. X.; Duan. S.; Xu. S.; Zhou. C.; *Electrochim. Acta* 2011. *58*. 142.

Yin. H.; Zhang. Q.; Zhou. Y.; Ma. Q.; Iiu. T.; Zhu. L.; Ai. S.; *Electrochim. Acta* 2011. *56*. 2748.

Yuan. D.; Chen. S.; Yuan. R.; Zhang. J.; Liu. X.; *Sens. Actuators. B* 2014. *191*. 415.

Zhang. T.; Lang. Q.; Yang. D.; Li. L.; Zeng. L.; Zheng. C.; Li. T.; Wei. M.; Liu A.; *Electrochim. Acta* 2013. *106*. 127.

Zhang. Y.; Zheng. J. B.; *Electrochim. Acta* 2007. *52*. 7210.

Zhao. G.-H.; Tang. Y.-T.; Liu. M.-C.; Lei. Y.-Z.; Xiao. X.-E.; *Chin. J. Chem.* 2007. *25*. 1445.

Zheng. L.; Xiong. L.; Li. Y.; Xu. J.; Kang. X.; Zou. Z.; Yang. S.; Xia. J.; *Sens. Actuators. B* 2013. *177*. 344.

In: New Developments in Analytical Chemistry ... ISBN: 978-1-63463-427-4
Editor: Breanna Granger

Chapter 4

ADSORPTION PROPERTIES OF A NEW ION-SELECTIVE HIERARCHICALLY IMPRINTED POLY(ACRYLAMIDE-CO-EGDMA) FOR THE EXTRACTION/PRECONCENTRATION OF CO(II) FROM AQUEOUS MEDIUM

***Fernanda Midori de Oliveira*[1], *Nathália Barbosa Wutke*[1], *Affonso Celso Gonçalves Jr.*[2], *Mariana Gava Segatelli*[1] and *César Ricardo Teixeira Tarley*[*1,3]**

[1]Universidade Estadual de Londrina (UEL),
Departamento de Química, Centro de Ciências Exatas,
Rodovia Celso Garcia Cid, Londrina, PR, Brazil
[2]Universidade Estadual do Oeste do Paraná, Centro de Ciências Agrárias,
Marechal Cândido Rondon-PR, Brazil
[3]Instituto Nacional de Ciência e Tecnologia (INCT) de Bioanalítica,
Universidade Estadual de Campinas (UNICAMP),
Instituto de Química, Departamento de Química Analítica,
Cidade Universitária Zeferino Vaz, s/n, Campinas, SP, Brazil

[*] Corresponding author. Tel +55 43 3371 4366; fax +55 43 3371 4286; E-mail address: tarley@uel.br (C. R. T. Tarley).

ABSTRACT

The present study focuses on the synthesis of ion imprinted poly(acrylamide-co-EGDMA) using cobalt and cetyltrimethylammonium bromide (CTAB) as double templates and its application for extraction/preconcentration of Co(II) ions from aqueous medium. The preconcentration properties toward cobalt ions for the IIP-CTAB (hierarchically imprinted organic polymer) and NIP-no CTAB (without imprinting process) was evaluated and their characterization was accomplished by SEM, FT-IR, elemental analysis and TGA. IIP-CTAB provided better recognition capacity toward cobalt ions than that one achieved for other metal ions. From breakthrough curve a maximum adsorption capacity of 0.1496 mg g^{-1} for IIP-CTAB was obtained. The analytical feasibility of IIP-CTAB as adsorbent of cobalt ions was evaluated by the on-line coupling of a preconcentration flow injection system to FAAS optimized by means of experimental design methodology. The preconcentration procedure was based on sample loading at pH 8.65 (20.0 mL) through 50.0 mg of IIP-CTAB packed into a mini-column at a fast flow rate of 9.0 mL min^{-1} with further elution with 2.5 mol L^{-1} HNO_3. The linear dynamic range for cobalt was from 5 up to 200.0 μg L^{-1} with limit of quantification (LOQ) of 0.59 μg L^{-1}, preconcentration factor (PF) of 37 and high sample throughput of 22 h^{-1}. The sensitivity of method using IIP-CTAB as adsorbent was 1.75 times higher than one achieved for NIP-no CTAB. Experimental results showed that adsorbent was very stable chemically since it supported more than 200 cycles of preconcentration and desorption using 2.5 mol L^{-1} HNO_3 as eluent. The preconcentration method was used to determine cobalt in water samples.

Keywords: Surfactant, ion imprinted polymer, FAAS, breakthrough curve

1. INTRODUCTION

Cobalt (Co) is considered an essential micronutrient for human body, being a component of cyanocobalamin or vitamin B_{12} that is responsible to red blood cells synthesis (Munõz et al., 2002). On the other hand, excessive levels of cobalt in body human can induces cytoxicity of human keratinocytes, reproductive problems, hypertension, hyperglycemia and pulmonary diseases (Baruthio et al., 2007; Alibabic et al., 2007). As cobalt has been widely used in many industrial processes of alloys, pigments, catalysts and batteries the major contamination source of this element occurs in aquatic body with consequent contamination of aquatic organisms, especially by bioaccumulation in fishes (Alibabic et al., 2007). Therefore, the removal or monitoring of cobalt in aquatic bodies is of paramount importance. The cobalt monitoring usually requires the implementation of preconcentration methods due to the low concentration of element in aquatic body in combination with a high percentage of interfering matrix components. There exist various processes aimed at removal/preconcentration cobalt-contaminated water including liquid-liquid extraction (LLE) (Okamoto et al., 2000), dispersive liquid-liquid microextraction (DLLME) (Amirkavei et al., 2013), coprecipitation (Soylak et al., 2007), cloud point extraction (Citak et al., 2010), liquid membrane separation (Sadyrbaeva et al., 2013) and solid phase extraction (SPE) (Ghaedi et al., 2007), but this latter one is gaining preeminence due to its low cost, simplicity, selectivity depending upon nature of adsorbent, ease of automation, ease of operation, as well as not causing the formation of noxious substances (Ferreira et al., 2007). Numerous adsorbents have been developed for cobalt and some of them can be cited: carbon nanotubes (Souza et al., 2009), polyurethane foam loaded with Azo compounds (Matos et al., 2005), mixed oxides (Diniz et al., 2014), modified Amberlite XAD resins (Kumar, 2000), Chelex 100 (Nomngongo et al., 2013), Diaion SP-850 resin (Soylak et al., 2006) and ion imprinted polymers (IIP) (Marestoni et al., 2013). Among this huge range of adsorbents, IIP have been studied as potential selective adsorbents, because it exploits the molecular recognition technology, where functional and cross linking monomers are co-polymerized in the presence of the metallic ion (template) with results in trapping of template in polymeric matrix. Removal of the template from polymeric matrix renders selectivity cavities complementary in shape and size to that of the template able to selectively re-adsorb the template (Marestoni et al., 2013). A survey of literature demonstrates that some approaches have been developed for preparation of cobalt imprinted adsorbent, such as surface-imprinting polymerization (Tsukagoshi et al., 1993), bulk polymerization using diazoaminobenzene (DAAB) and vinylpyridine (VP) as monomers (Tajodini et al., 2010), bidimensional ion imprinted adsorbent (Tarley et al., 2011), mesoporous silica SBA-15-supported surface ion imprinted polymer (Liu et al., 2011) and more recently ion-imprinted mesoporous silica hybrids (Moorthy et al., 2013). This latter approach fundamentally exploits the double imprinted concept as named hierarchically imprinted process, where makes use of ionic and micellar imprinting technology able to create hierarchical structures at different scales (Dai 2001). Therefore, the resulting material containing the selective binding sites presents easier access toward analyte, a more favorable reversibility and a fast adsorption/desorption process; such features are highly desirable in column solid phase extraction system on-line coupled to flow injection analysis. Even though considering the advantages of these materials, at the best of our knowledge, there are no researches devoted to the development of new preconcentration methods for Co(II) based on adsorbent flow injection analysis for metal ions using hierarchically imprinted polymers. Furthermore, the most hierarchical imprinted polymer are synthesized through the sol-gel process, whose adsorption studies have been carried out in batch or in open column (off-line) for the metals Hg(II), Cd(II) and Ni(II) (Wu et al., 2007; Lu et al., 2004; Nacano et al., 2010).

In this context, this work deals with synthesis and characterization of new ion-selective hierarchically imprinted poly(acrylamide-*co*-ethyleneglycol-dimethacrylate) and its separation properties for Co(II) from aqueous medium. The hierarchical imprinted organic copolymer was characterized by FT-IR, SEM, TGA and its analytical feasibility as adsorbent for Co(II) was evaluated by developing a flow injection system on-line coupled to FAAS. The attractive features of hierarchically imprinted organic polymer for flow injection analysis although even a few studied has been recently reported by our research group who synthesized a cross-linked poly(methacrylic acid-*co*-ethyleneglycoldimethacrylate) for Ni(II) preconcentration in water samples (Oliveira et al., 2014).

2. EXPERIMENTAL

2.1. Reagents

Ethylene glycol dimethacrylate (EGDMA) as cross-linking reagent and 2,2'-azobisisobutyronitrile (AIBN) as radical initiator were purchased from Sigma-Aldrich (Steinheim, Germany) and used without prior purification. Ethanol (CH_3CH_2OH), nitric acid (HNO_3), sodium hydroxide (NaOH), cobalt nitrate hexahydrate ($Co(NO_3)_2.6H_2O$) and sodium dihydrogen phosphate (NaH_2PO_4) were purchased from Merck (Darmstadt, Germany). Cetyl-trimethylammonium bromide (CTAB), tris(hydroxymethyl) aminomethane (Tris-HCl) and the monomer acrylamide were acquired from Acros (New Jersey, USA), Invitrogen (Carlsbad, USA) and Vetec (Rio de Janeiro, Brazil), respectively. Cobalt standard solutions were prepared from standard stock solution of 1000 mg L^{-1} Reagenti Carlo Erba (Rodano, Italy) with appropriate dilutions. All solutions were prepared using water purified by the Milli-Q® system (Bedford, USA).

2.2. Apparatus

The cobalt measurements were carried out using a Shimadzu AA-6601 flame atomic absorption spectrometer (Shimadzu, Tokyo, Japan) equipped with a cobalt hollow cathode lamp (Hamamatsu Photonics KK) operated with a current of 12 mA and a wavelength of 240.7 nm, as radiation source and a deuterium lamp for background correction. Acetylene and air were supplied to the burner at 2.2 and 15.0 L min^{-1}, respectively. For the on-line preconcentration system, the standard solution, sample and eluent were propelled by using Ismatec peristaltic pump (Model IPC (IPC Ismatec-08 Glattzbrugg, Switzerland)) containing Tygon® tubing. The steps of preconcentration elution were carried out by using a home-made injector commutator made of poly(methylmetacrylate). Polyethylene tubes 0.8 mm in diameter was used to carry the sample and eluent. A spectrometer model Shimadzu FT-IR 8300 operating in transmission mode between 4000 cm^{-1} and 400 cm^{-1} was used to elucidate functional groups present in the adsorbent by use of KBr pellets. The measurements of pH were performed using a pH meter Metrohm 827 pH lab. For data acquisition related to the analysis of variance, the program STATISTICA (version 6.0) was used. The polymers morphology was evaluated by Scanning Electron Microscopy (SEM) using a JSM-300LV. The polymers were coated with a thin layer of gold using a Bal-Tec MED 020 equipment, in order to minimize charging under the incident electron beam irradiation. Thermal stability of the sample was evaluated by thermogravimetry (TG), using a Perkin Elmer TGA 4000 (Massachusetts, USA) in the temperature range of 30–700 °C, at a scanning rate of 20 °C min^{-1}, under flowing nitrogen (20 mL min^{-1}). Carbon, nitrogen and hydrogen contents in the synthesized materials were determined using a Perkin Elmer 2400 elemental analyzer (Waltham, MA, USA).

2.3. Synthesis of Polymers

The synthesis of poly(acrylamide-*co*-EGDMA) was performed in a round-bottom flask, whereas 0.2096 g of cetyltrimethylammonium bromide (CTAB) was dissolved in 10.0 mL of ethanol, followed by addition of 0.1023 g of

$Co(NO_3)_2.6H_2O$, and 1.2537 g of acrylamide. The obtained mixture was mixed manually with a glass rod. Then, 250 mg of 2,2'-azobisisobutyronitrile (AIBN) and 5.0 mL of ethylene glycol dimethacrylate (EGDMA) were added to the mixture. Nitrogen gas was bubbled for 5 min and then the flask was sealed. This mixture was polymerized at 60 °C for 24 hours. Subsequently, the polymer was removed from the flask and separated into particles smaller than 106 μm. Finally, the material was washed with absolute ethanol to remove the surfactant and 2.5 mol L^{-1} HNO_3 solution for removal the analyte (Co^{2+}). The blank polymer NIP-no CTAB (without imprinting process) was synthesized in similar way. The IIP-CTAB was packed into a mini-column of polyethylene (4.2 cm x 1.0 cm i.d.). In the edges of mini-column cotton was added to prevent loss of material.

2.4. Multivariate Optimization of the On-Line Coupled Preconcentration System

The on-line coupled preconcentration system to FAAS is affected by several variables, such as sample pH, buffer concentration, eluent concentration, mass of adsorbent packed into mini-column and preconcentration flow rate. Therefore, in order to assess the effect of these variables on the cobalt determination, a 2^{5-1} fractional factorial design (first-order design) was carried out. The values corresponding to high (+) and low (-) levels for each variable are shown in Table 1.

This chemometric tool enables to reduce the number of experiments and help to avoid trivial mistakes during optimization in comparison to one-way design. However, although the 2^{5-1} fractional factorial design does not lead to final optimization of method, its application makes possible to obtain the direction of experimentation to be taken into account in successive optimization stages to find the final optimization. In this sense, after screening variables, a second-order design, Doehlert matrix was employed for the final optimization with those significant variables. The analytical response was taken as absorbance (peak height) and the experiments were carried out by percolating 20.0 mL of a cobalt standard solution at 200 μg L^{-1} concentration. Statistical analysis (analysis of variance – ANOVA with confidence interval of 95%) of obtained data and graphically represented by Pareto chart were performed with the STATISTICA 6.0 software package (Tulsa, USA).

Table 1. Factors, their levels and 2^{5-1} fractional factorial design

Factors						High (+)	Low (-)
pH						9.0	7.0
Buffer Concentration (BC) [mol L^{-1}]						0.10	0.01
Eluent Concentration (EC) [mol L^{-1}]						2.5	1.5
Mass of adsorbent (MA) [mg]						100	50
Preconcentration flow rate (PFR) [mL min^{-1}]						9.0	5.0
Assays	pH	BC	EC	MA	PFR	Absorbance	
1	1	1	1	1	1	0.0763	0.0715
2	1	1	1	-1	-1	0.0598	0.0639
3	1	1	-1	1	-1	0.0776	0.0813
4	1	1	-1	-1	1	0.0578	0.0585
5	1	-1	1	1	-1	0.2039	0.2061
6	1	-1	1	-1	1	0.2427	0.2518
7	1	-1	-1	1	1	0.1390	0.1384
8	1	-1	-1	-1	-1	0.2449	0.2451
9	-1	1	1	1	-1	0.0381	0.0357
10	-1	1	1	-1	1	0.0266	0.0256
11	-1	1	-1	1	1	0.0363	0.0385
12	-1	1	-1	-1	-1	0.0300	0.0282
13	-1	-1	1	1	1	0.0496	0.0472
14	-1	-1	1	-1	-1	0.0515	0.0486
15	-1	-1	-1	1	-1	0.0499	0.0560
16	-1	-1	-1	-1	1	0.0521	0.0532

2.5. On-line Preconcentrion Flow Injection System Coupled to FAAS

Through the mini-column containing 50.0 mg of IIP-CTAB, 20.0 mL of sample, buffered with 0.007 mol L^{-1} ammoniacal buffer (pH 8.65) at a flow rate of 9.0 mL min^{-1} were percolated. The elution step was accomplished by using 2.5 mol L^{-1} HNO_3 was continuously pumped into the detector FAAS a flow rate of 5.0 mL min^{-1}, the same flow rate of FAAS nebulizer. After preconcentration step, the elution step was performed by switching the central part of home-made injector commutator, whereas the cobalt ions were continuously eluted with 2.5 mol L^{-1} HNO_3 toward FAAS detector at a flow rate of 5 mL min^{-1}. The schematic diagram of the on-line coupled preconcentration system to FAAS is depicted in Figure 1.

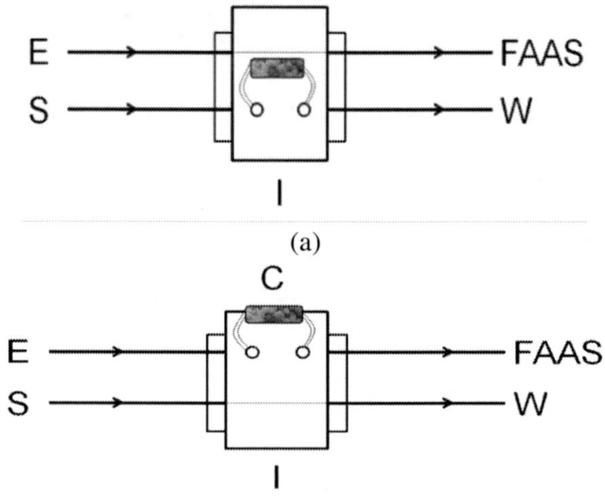

Figure 1. Schematic diagram of on-line coupled preconcentration system to FAAS. (a) preconcentration mode, and (b) elution mode. S- sample solution; E- eluent; I- injector/commutator; C- mini-column with the adsorbent; W- waste; FAAS- spectrometer.

2.6. Selectivity Studies

The selectivity studies were carried in triplicate using batch procedure. For these experiments 180 mg of each polymer (IIP-CTAB and NIP-no CTAB) were placed in 200.0 mL cylindrical polystyrene flasks containing 25.0 mL of 2.5 mg L^{-1} binary solutions [Co(II)/Cu(II), Co(II)/Zn(II), Co(II)/Cd(II)] buffered with 0.007 mol L^{-1}ammoniacal (pH 8.65) and stirred for 30 min. The intervening ions were chosen due to their similarity in charge and size (ionic radius Co(II) = 70 pm, Cu (II)= 73 pm, Zn(II)= 74 pm, Cd(II)= 95 pm) and the ability to interact well with nitrogen and oxygen atoms present in the acrylamide structure. At the stirring time, the supernatant was centrifuged and the concentrations of cobalt remaining and the intervening ions were determined by FAAS. From the initial ($C_{initial}$mg L^{-1}) and final (C_{final}mg L^{-1}) concentration the distribution coefficient (K_d) was calculated as equation (1):

$$K_d = \frac{(C_{initial} - C_{final})}{C_{final}} \frac{V}{M} \tag{1}$$

where V is the solution volume (mL); M is the mass of polymer (g) of the IIP-CTAB or NIP-no CTAB. The selectivity coefficient *k* (dimensionless) obtained for IIP-CTAB and NIP-no CTAB by competitive adsorption for cobalt ion in the presence of competitor ion was calculated by equation (2):

$$k = \frac{k_d(Co)}{K_d metal} \tag{2}$$

The relative selectivity (k') coefficients were than determined according to equation (3):

$$k' = \frac{k_{IIP-CTAB}}{k_{NIP-noCTAB}} \tag{3}$$

2.7. Breakthrough Curve

In order to determine the loading dynamic capacity of the IIP-CTAB, a breakthrough experiment was performed. For this task, the same mini-column packed with 50.0 mg of IIP-CTAB used in the flow injection analysis was submitted to loading with aliquots of 1.0 mL of 0.633 mg L^{-1} Co(II) solution buffered with 0.007 mol L^{-1} ammoniacal buffer (pH 8.65) at a flow rate of 9.0 mL min^{-1} until saturation was reached. Each aliquot from mini-column effluent was further collected and the amount of Co(II) (C_{final}) was determined by FAAS. From these results a plot $C_{final}/C_{initial}$ *vs.* effluent volume (mL) was obtained making it possible to obtain the breakthrough volume and the maximum adsorption capacity under dynamic condition, when $C_{final}/C_{initial} \cong 0$ and $C_{final}/C_{initial} \cong 1$, respectively.

3. Results and Discussion

3.1. Characterization of the Polymers

The FT-IR spectra of the polymers are shown in Figure 2. The signals at 3444 and 1669 cm^{-1} are attributed to stretching and bending of –OH groups from water molecules adsorbed. The first band one can also be overloaded with NH_2 stretching and vibration absorptions, and the second one with C=C stretching of vinyl groups from EDGMA and stretching vibration of C=O and N-H bending, which are the $CONH_2$ group (Oliveira et al., 2013; Segatelli et al., 2010; Sahiner et al., 2005; Xu et al., 2011; Merlin et al., 2009). The absorptions at 2998 and 2956 cm^{-1} are related to the C-H asymmetric stretching of $-CH_2-$ and $-CH_3-$ in the polymer chain and EGDMA, respectively. The signal at 1729 cm^{-1} can be observe the stretching of the C=O bond present in EGDMA and acrylamide. The absorption at 1459 cm^{-1} is related to the angular deformation of CH_2 (scissors). The angular deformation out of the plane of the methyl group may be observed at 1385 cm^{-1} (umbrella). The signals at 1262 and 1156 cm^{-1} are assigned to C-O stretching of acrylamide and EGDMA overlap with stretching C-N (Orozco-Guareño et al., 2010). From the results obtained by FT-IR, it can be suggested that the functional groups had been anchored in the polymeric network.

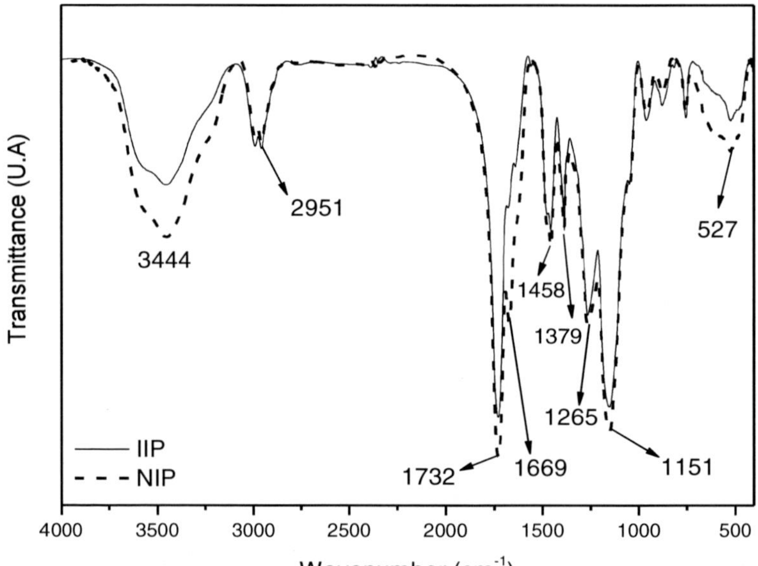

Figure 2. FT-IR spectra of IIP-CTAB and NIP-no CTAB.

SEM images of polymers are shown in Figure 3. As observed the polymers presented well-defined spherical particles and uniform size distribution. The particles aggregates resemble a bunch of grapes owing to the high hydrophilic surface and a low density of materials; such features provide good conditions for on-line solid-phase extraction with quick adsorption/desorption kinetics.

In general, it could be seen that particles size of IIP-CTAB were found to be lower, more defined and with lower aggregation degree in comparison to NIP-no CTAB. Such a result can be attributed to the higher solubility of cobalt-acrylamide complex in the presence of CTAB into chain growth reactions of poly(acrylamide-*co*-EGDMA) as well as by lower solubility in the porogenic solvent (Renkecz et al., 2013).

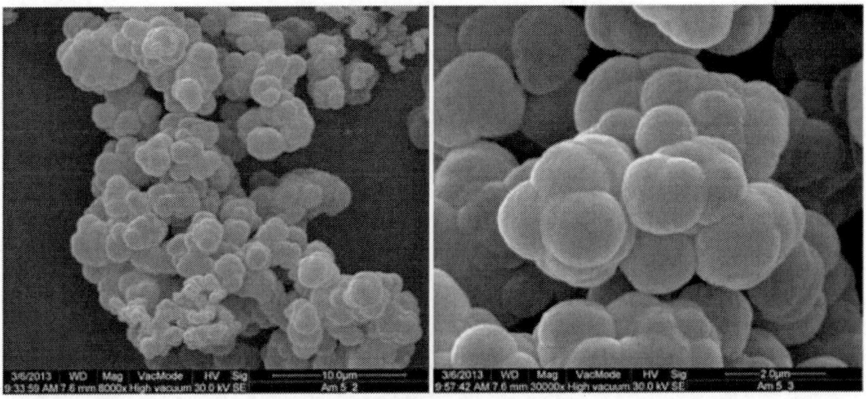

(a)

(b)

Figure 3. SEM images of(a) IIP- CTAB (b) NIP-no CTAB. The images were amplified in 8000 and 30000 times.

Thermogravimetric analysis shown in Figure 4 was useful in obtaining enough information on the polymer relative thermal stability. As observed from Figure 4, at about 35-100 °C, the polymers IIP-CTAB and NIP-no CTAB supports, respectively, 7.4 and 12.8% weight losses, which is attributed to the removal of physically adsorbed water. During the second stage of heating (300-540°C), a more pronounced weight loss was observed being, respectively, of 81.4 and 82.5 % for IIP-CTAB and NIP-no CTAB, which can be ascribed to the overlap of decrosslinking of two networks structures formed by degradation of poly(acrylamide) and poly(EGDMA) segments (Merlin et al., 2009; Shao et al., 2010). The results also demonstrate that NIP-no CTAB supports higher temperature without an accentuated weight loss as regards IIP-CTAB, as observed, for instance, at 350 °C. Such a result makes possible to stress out that the surfactant template (CTAB) leads to a decrease in the thermal behavior of IIP-CTAB probably due to pores creation in the network and, as a consequence, an decrease in the crosslink density. These results can also corroborate the SEM data, in which NIP-no CTAB presents a higher aggregation degree in comparison to IIP-CTAB.

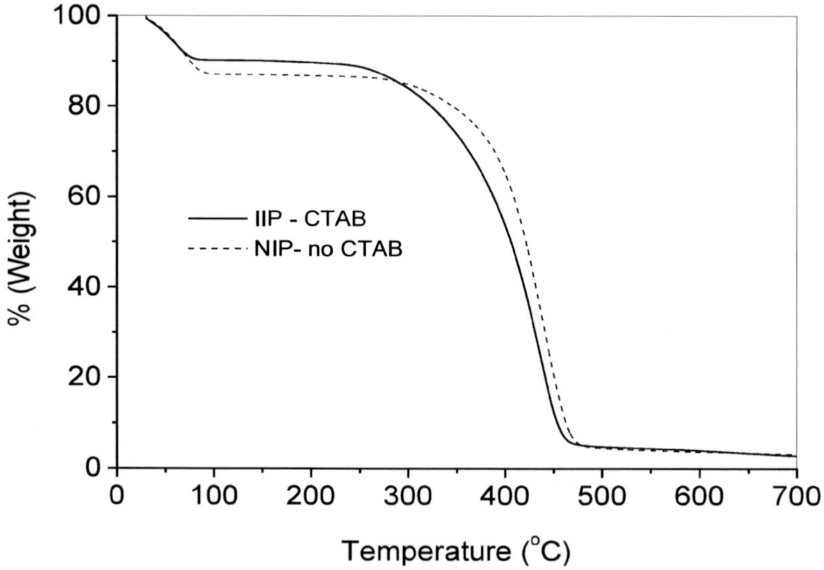

Figure 4. TGA of IIP-CTAB and NIP-no CTAB.

The monomer reactivity ratios were evaluated from elemental analysis. The content of C, H and N for IIP-CTAB and NIP-no CTAB were found to

52.97%, 7.46%, 2.25% and 52.05%, 7.56%, 2.32% respectively. The molar percentage of monomers acrylamide and EGDMA (in mol %) used in the synthesis, for both copolymers, were found to be 40.39% and 59.61% for acrylamide and EGDMA, respectively. From elemental analysis and using stoichiometric calculation the respective copolymer composition (in mol%) for IIP-CTAB and NIP-no CTAB were found to be 66.76%, 33.24% and 67.65%, 32.35%. As observed from these data, the final molar percentage of acrylamide (functional monomer) in both copolymers were higher than that one used in the synthesis, thus indicating a better conversion of this monomer unit in the copolymer composition. In addition, for the copolymers herein studied there is no evidence that surfactant changes the reactivity of monomers intochain growth of poly(acrylamide-*co*-EGDMA), but only the solubility of cobalt-acrylamide as before mentioned from SEM data.

3.2. Multivariate Optimization of the On-Line Coupled Preconcentration System

The preliminary study from the 2^{5-1} fractional factorial design, whose effects were graphically represented by Pareto chart (Figure 5), shows that with increasing sample pH (7.0 up to 9.0), absorbance increases. The sample pH not only impacts the surface structure of adsorbents and the metal ion formation but it plays an important role in the interaction between adsorbent and metal ion. Therefore, such result suggest that at pH 9.0 the adsorption of cobalt was much higher and may be rationalized on the basis of the deprotonation of nitrogen atoms present in the acrylamide monomer, at which the cobalt adsorption becomes more favorable through complex formation or electrostatic interaction. Buffer concentration (BC) was the second more important variable in the system but with negative effect (-88.03), which indicates that a high buffer concentration, the adsorption of cobalt is diminished. It could be emphasizing that the buffer concentration x pH interaction was statistically significant with negative effect (-67.4), showing clearly that these variables cannot be optimized by one-way design.

The mass of adsorbent (MA) has a significant negative effect (-13.4) on the response. Such result can be explained due to great broadening of the analytical signal owing to dispersion of sample zone into the mini-column, and as consequence, a decrease of analytical signal. Therefore, the mass of adsorbent of 50 mg was adopted in the future experiments.

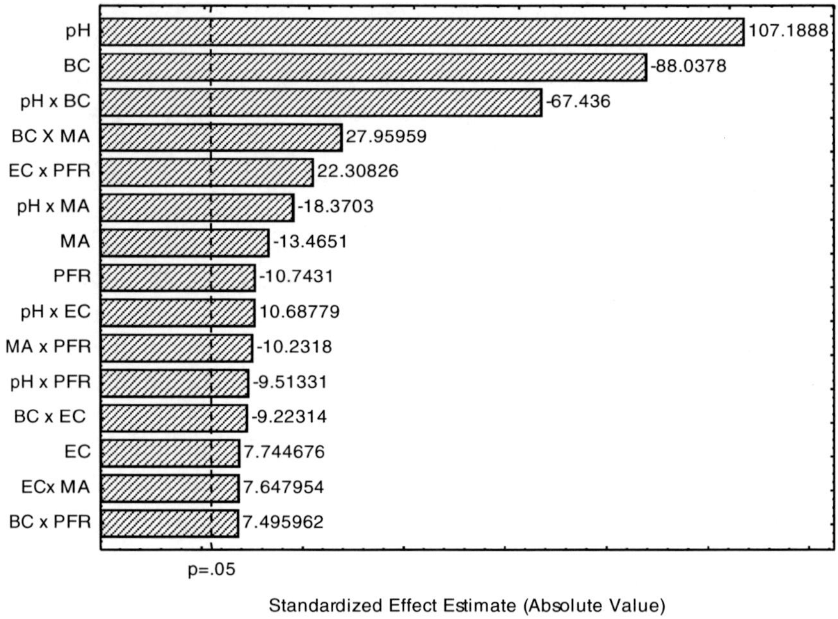

Figure 5. Pareto chart of standardized estimated effects caused by the factors tested in the 2^{5-1} fractional factorial design.

Table 2. Doehlert design employed for the pH and BC optimization

Assays	pH	BC	Absorbance	Buffer solution type
1	0 (8.50)	0 (0.01)	0.3150	Ammoniacal
2	0 (8.50)	0 (0.01)	0.3251	Ammoniacal
3	0 (8.50)	0 (0.01)	0.3257	Ammoniacal
4	0 (8.50)	0 (0.01)	0.3400	Ammoniacal
5	1 (10.50)	0 (0.01)	0.1801	Ammoniacal
6	0.5 (9.50)	0.866 (0.1)	0.0899	Ammoniacal
7	-1 (6.5)	0 (0.01)	0.1101	Phosphate
8	-0.5 (7.5)	-0.866 (0.001)	0.2798	Tris-HCl
9	0.5 (9.5)	-0.866 (0.001)	0.2809	Ammoniacal
10	-0.5 (7.5)	0.866 (0.1)	0.0811	Tris-HCl

The preconcentration flow rate (PFR) investigated under experimental domain (5.0 to 9.0 mL min^{-1}) provided a slight negative effect (-10.7), which implies that higher values of PFR decrease somewhat the analytical signal due to lower mass transfer of cobalt towards the polymer surface. However, in order to assure a satisfactory sample throughput of the on-line preconcentration system, the PFR was set at 9.0 mL min^{-1} and further experiments were not performed for final optimization of PFR.

The eluent concentration (EC) was set to the highest level (2.5mol L^{-1}) due to its positive effect value obtained (7.7). This concentration was more efficient to elute cobalt ions without memory effect in approximately 30 sec.

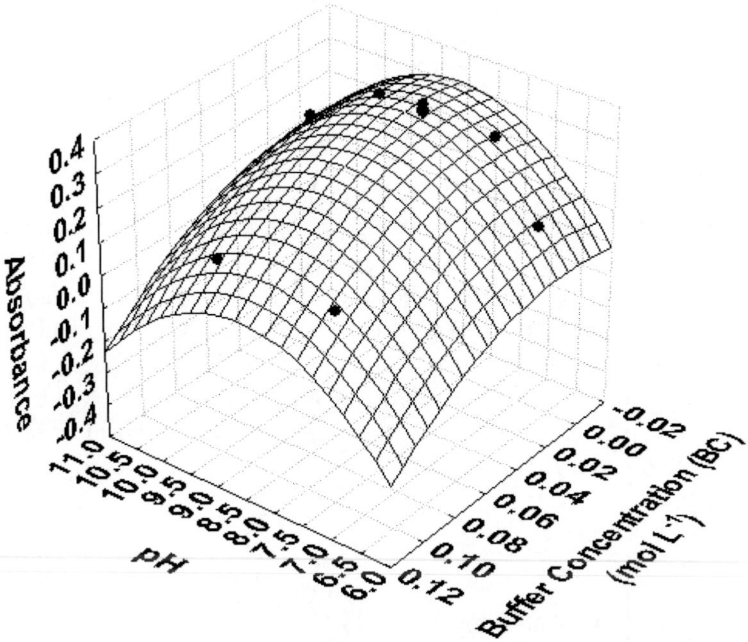

Figure 6. Response surface constructed for the interaction between the factors (pH and BC).

According to aforementioned, the final optimization of pH and buffer concentration was carried out by using a Doehlert matrix with quadruplicate at central point. For this design, the levels were chosen, according to preliminary study from the 2^{5-1} fractional factorial design. Table 2 shows the values which correspond to the levels for each factor and as can be observed, the best analytical signal was obtained at central point. The fitted second-degree equation obtained from the Doehlert matrix within experimental domain was found to be Abs= $-3.073 + 0.785pH - 0.0453pH^2 + 1.080CT - 22.810CT^2 - 0.0875pH*CT$. The analysis of the results from the ANOVA with confidence interval of 95% showed that this model is adequate to model the experimental data because the lack of fit was not significant [$F_{calculated}(4.80) < F_{tabled}(7.71)$]. From the second-degree equation a surface response was constructed, showing a maximum point at pH 8.65 and ammoniacal buffer concentration of 0.007 mol L^{-1} (Figure 6).

Therefore, the final suitable experimental condition for the on-line preconcentration method of cobalt was found to be as pH 8.65, 0.007 mol L^{-1} ammoniacal buffer, preconcentration flow rate of 9.0 mL min^{-1}, 2.5 mol L^{-1} HNO_3 as eluent and mass polymer of 50 mg.

3.3. Selectivity Studies

The ionic imprinting on the ion imprinted polymers depends on the affinity of the ligand (functional monomer) with metal ion, on the number and type of interaction sites and charges and sizes of metal ion. So, a higher imprinting effect can be expected when the template shape and the monomer–template rigidity with the functional monomers are likely to yield binding sites. Acrylamide-based adsorbent has been employed as efficient adsorbent for removal of toxic metal ions from aqueous medium with higher affinity towards Hg(II), Pb(II), Cd(II) but with a relatively lower affinity for Co(II) (Zhou et al., 2011; Kesenci et al., 2002).

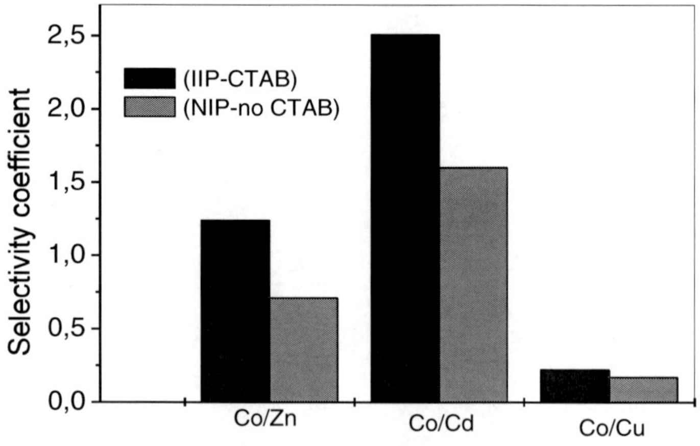

Figure 7. Selectivity coefficient of Co(II), Zn(II), Cd(II), and Cu(II) onto IIP-CTAB and NIP-no CTAB in binary mixture.

In this sense, it would expected a lower selectivity coefficient toward Co(II); however, as observed from the Figure 7, the competitive adsorption of binary metal ions Co(II)/Ni(II), Co(II)/Zn(II), Co(II)/Cd(II) and Co(II)/Cu(II) revels that the selectivity coefficient (*k*) obtained for IIP-CTAB was higher than that one achieved for NIP-no CTAB. These results reflect in relative selectivity coefficients (*k´*) greater than 1, indicating a better selectivity toward Co(II) over other metal ions when using IIP-CTAB and confirming the successful imprinting effect on the hierarchically imprinted poly(acrylamide-*co*-EGDMA).

3.5. Adsorption Capacity from Breakthrough Curve

Figure 8 depicts the breakthrough curve. The condition $C_{final}/C_{initial} \cong 0$ was achieved when using 4.0 mL of 0.633 mg L^{-1}Co(II) solution buffered with 0.007 mol L^{-1}ammoniacal buffer (pH 8.65) at a flow rate of 9.0 mL min^{-1}. Under this condition the breakthrough capacity was found to be 0.051 mg g^{-1}.

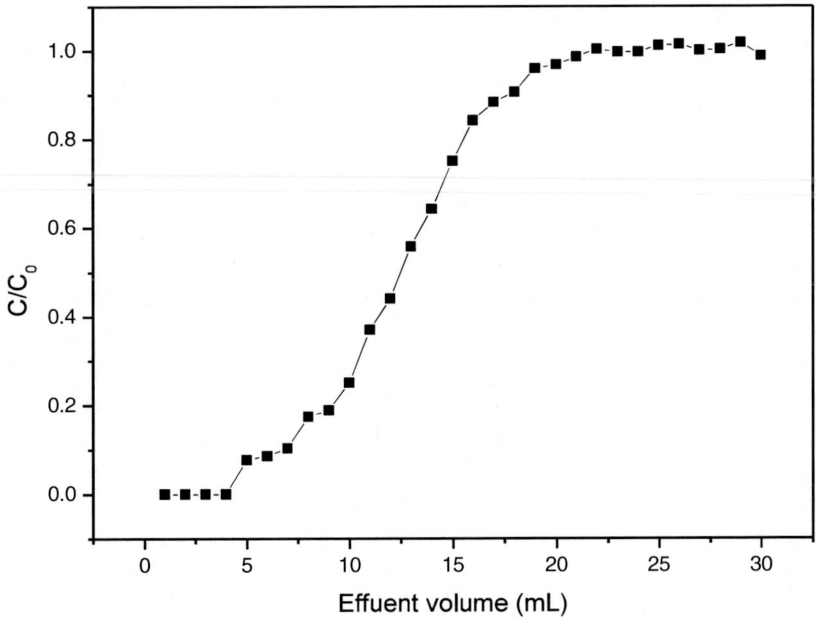

Figure 8. Breakthrough curve for Co(II) under dynamic conditions.

The maximum adsorption capacity under dynamic condition was achieved when $C_{final}/C_{initial} \cong 1$ by loading 24.0 mL of Co(II) standard solution yielding saturation of binding sites of copolymer at 0.150 mg of Co(II) per gram of polymer. The value of breakthrough capacity plays an important role in improving the detectability of preconcentration method. From the obtained value (0.051 mg g^{-1}), we can assure that the method can support a higher sample loading than that one employed (20.0 mL) without losses of adsorption capacity, thus yielding lower limits of detection and quantification.

3.6. Interference Studies, Analytical Performance of the On-Line Preconcentration Method and Application

The influence of selected metal ions on the cobalt preconcentration under optimal experimental condition was evaluated. Binary solutions containing Co(II) with Na(I), K(I), Ba(II), Zn(II), Cu(II), Ni(II), Mn(II), Cr(III), Cd(II) or Pb(II) at difference analyte:interfering ion ratios were submitted to the preconcentration system and the analytical signals were compared to that one achieved by a single solution containing Co(II). The tolerance limit for interfering ions was considered as the maximum concentration that gave a relative error less than ± 10 % at a concentration of 100 μg L^{-1} of Co(II). Ni(II), Cu(II), Zn(II), Pb(II) and Mn(II) did not interfere until a 50-fold excess was achieved. Cd(II) and Cr(III) showed no changes on analytical response of cobalt even a 10-fold excess. Finally, even a 100-fold excess of Na(I), K(I) and Ba(II) no changes on cobalt signal was observed. Such results suggest that cobalt determination in water samples was not affected by concomitant ions.

The linearity of method was studied for cobalt concentration ranging from 0 (blank) up to 200.0 μg L^{-1} providing a linear equation of Abs= 0.00182[Co(II)] + 0.00369 with a correlation coefficient of 0.999. The linear equation obtained by direct aspiration of cobalt solution without preconcentration step was found to be Abs = 0.00004912[Co(II)] + 0.00608 with a correlation coefficient of 0.99846. From the ratio between the slopes of these linear equations a high preconcentration factor of 37-fold. The limits of detection (LOD= 0.17 μg L^{-1}) and quantification (LOQ= 0.59 μg L^{-1}) calculated according to the IUPAC definition (Long et al., 1983) as three and ten times, respectively, the relative standard deviation (RSD) of ten blank measurements divided by the slope of the calibration curve constructed for the preconcentration procedure. The concentration efficiency (CE) obtained as 16.6 min^{-1} was calculated as the preconcentration factor achieved in 1 min during preconcentration step. In the present study 20.0 mL of sample at a flow rate of 9.0 mL min^{-1} were percolated though mini-column of poly(acrylamide-*co*-EGDMA) during 2.2 min. The consumptive index (CI), which represents the ratio of the preconcentration volume (20.0 mL) to the PF, was found to be 0.58 mL. The sample throughput of proposed method was found to be 22.6 h^{-1} taking into accounts the times, 132 and 27 sec employed in the preconcentration and elution steps.

In order to evaluate the improvement of dual imprinted effect on the sensitivity of proposed preconcentration method using the poly(acrylamide-*co*-EGDMA), a calibration plot for the Co(II) using NIP-no CTAB was constructed. A linear equation Abs= 0.00104[Co(II)]+0.00178 (r= 0.996) was obtained, whose sensitivity was about 1.75-fold lower than that one achieved using IIP-CTAB, revealing the good adsorption of IIP-CTAB with respect to NIP-no CTAB.

The proposed method was applied in waters samples. The mineral water was acquired in a local supermarket and the tap water was obtained in the Londrina State University. The river and lake waters was collected in bottles with subsequent acidification with nitric acid in order to prevent the growth of microorganisms, after that this waters was filtered in Nylon® membranes. In all samples the cobalt concentration was below the detection limit of the method. However when to realize the samples spike satisfactory recovery percentages was obtained which ranged from 91.1 to 107.8%.

Table3. Recovery percentages of Co(II) after spike in waters samples

Sample	Co(II) Ions added (μg/L)	Co(II) concentration found (μg/L)	Recovery (%)
River water	0	ND*	-
	10	10.78	107.8
Tap water	0	ND	-
	10	9.11	91.1
Mineral water	0	ND	-
	10	9.11	91.1
Lake water	0	ND	-
	10	9.67	96.7

*ND- Not detected.

Numerous solid phase preconcentration methods have been conducted for Co(II) determination by FAAS. Table 4 presents a comparative data on on-line and off-line solid phase preconcentration methods for Co(II) determination by

FAAS. It can be observed from table that the proposed method using the hierarchically imprinted poly(acrylamide-*co*-EGDMA) shows better analytical features with respect to other methods, especially regarding the Co(II)-imprinted diazoaminobenzene-vinylpyridine copolymer (Tajodini et al., 2010).

Table 4. Comparative data on on-line and off-line solid phase preconcentration methods for Co(II) determination by FAAS

Adsorbent	Preconcentration modality	Volume (mL)	FP	LD	Eluente	Referência
$SiO_2/Nb_2O_5/ZnO$	on-line	16	26	0.28	HCl	(Diniz et al., 2014)
Nano ZrO_2/B_2O_3	on-line	100	10	3.8	HNO_3	(*Yalçinkaya et al., 2011)*
5,7-dichloroquinone -8-ol embedded styrene–ethylene glycol dimethacrylate	off-line	1000	200	2.0	HCl	(Praveen et al., 2005)
L-tyrosine immobilized on multiwall carbon nanotubes	on-line	10	180	0.05	HNO_3	(Pacheco et al., 2009)
Functionalized Amberlite XAD-2 with pyrocatechol	on-line	5	23.0	1.98	HCl	(Lemos et al., 2006)
C_{18} functionalized with 1-nitroso-2-naphthol	on-line	3.25	17.2	3.2	Ethanol acidulated with 1% HNO_3	(Ye et al., 2002)
Co(II)-imprinted diazoaminobenzene-vinylpyridine copolymer	off-line	2000	200	0.05	HCl + thiourea	(Tajodini et al., 2010)
Hierarchically imprinted poly(acrylamide-*co*-EGDMA)	on-line	20	37.05	0.18	HNO_3	This work

CONCLUSION

The new-ion selective hierarchically imprinted poly(acrylamide-co-EGDMA) provided a high enrichment factor and low limits of detection and quantification for cobalt determination in water samples. From the competitive adsorption studies, the imprinted polymer presented higher selectivity for Co(II) in comparison to the non-imprinted polymer. Furthermore, the material showed a high reusability (more than 200 cycles of preconcentration-elution), thus indicating the high chemical and physical stability. For final remarks, it can be stated that the proposed preconcentration method using an ion imprinted polymer prepared through double imprinting process has characteristics that favor its use in routine analysis for the determination of cobalt in water samples.

ACKNOWLEDGMENTS

The authors would like to thank the Conselho Nacional de Desenvolvimento Científico e Tecnológico (CNPq), Fundação Araucária do Paraná, Coordenação de Aperfeiçoamento de Pessoal de Nivel Superior (CAPES) and Instituto Nacional de Ciência e Tecnologia de Bioanalítica (INCT) for their financial support and fellowships.

REFERENCES

Alibabic, V., Vahcic, N., Bajramovic, M., *Environ. Monit. Assess.,* 2007, 131, 349.

Amirkavei, M., Dadfarnia, S., Shabani, A.M.H., *Quim. Nova*, 2013,36, 63.

Baruthio, F., Pierre, F., Biol. *Trace Elem. Res*., 2007, 39, 21.

Citak, D., Tuzen, M., *Food Chem. Toxicol*., 2010, 48, 1399.

Dai, S., *Chem. Eur. J.*, 2001, 7, 763.

Diniz, K.M., Gorla, F.A., Ribeiro, E.S., Nascimento, M.B.O., Corrêa, R.J., Tarley, C.R.T., Segatelli, M.G., *Chem. Eng. J.*, 2014, 239,233.

Ferreira, S.L.C., Andrade, J.B., Korn, M.G.A., Pereira, M.G., Lemos,V.A., Santos, W.N.L., Rodrigues, F.M., Souza, A.S., Ferreira, H.S., Silva, E.G.P., *J. Hazard. Mater*., 2007, 145, 358.

Ghaedi, M., Ahmadi, M.S., *J. Hazard. Mater*., 2007, 147, 226.

Kesenci, K., Say, R., Debizli, A., *Eur. Polym. J.*, 2002, 38, 1443.
Kumar, M., Rathore, D.P.S., Singh, A.K., *Talanta*, 2000, 51, 1187.
Lemos, V.A., Gama, E.M., Lima, A.S., *Microchim. Acta*, 2006, 153, 179.
Liu, Y., Liu, Z., Dai, J., Gao, J., Xie, J., Yan, Y., *Chin. J. Chem.*, 2011, 29, 387.
Long, G.L., Winefordner, J.D., *Anal. Chem.*, 1983, *55, 712A.*
Lu, Y.-K., Yan, X.-P., *Anal. Chem.*, 2004, 76, 453.
Marestoni, L.D., Sotomayor, M.D.P.T., Segatelli, M.G., Sartori, L.R., Tarley, C.R.T., *Quím. Nova*, 2013, 36, 1194.
Matos, G. D.,Tarley, C. R. T.,Ferreira, S. L. C.,Arruda, M. A. Z., *Eclet. Quím.*, 2005, 30, 65.
Merlin, D.L., Sivasankar, B., *Eur. Polym. J.*, 2009, 45, 165.
Moorthy, M.S., Tapaswi, P.K., Park, S.S., Mathew, A., Cho, H.-J., Ha,C.-S., *Microporous Mesoporous Mater.*, 2013,180, 162.
Munõz, E., Palmero, S., Garciá-Garciá, M. A. *Talanta*, 2002, 57, 985.
Nacano, L.R., Segatelli, M.G., Tarley, C.R.T., *J. Braz. Chem. Soc.*, 2010, 21, 419.
Nomngongo, P.N., Ngila, J.C., Msagati, T.A.M., Moodley, B., *Phys. Chem. Earth*, 2013, 66, 83.
Okamoto, Y., Nomura, Y., Nakamura, H., Iwamaru, K., Fujiwara, T., Kumamaru, T., *Microchem. J.*, 2000, 65, 341.
Oliveira, F.M., Gonçalves Jr, A.C., Dragunski, D.C., Segatelli, M.G., Tarley, C.R.T.,Int. *J. Environ. Anal. Chem.*, 2014, 1.
Oliveira, F.M., Somera, B.F., Ribeiro, E.S., Segatelli, M.G., Yabe, M.J.S., Galunin, E., Tarley, C.R.T., *Ind. Eng. Chem. Res.*, 2013, 52, 8550.
Orozco-Guareño, E., Santiago-Gutiérrez, F., Morán-Quiroz, J.L., Hernandez-Olmos, S., Soto, V., Cruz, W., Manríquez, R., Gomez-Salazar, S., *J. Colloid Interface Sci.*, 2010, 349, 583.
Pacheco, P.H., Smichowski, P., Palla, G., Martinez, L.D., *Talanta*, 2009, 79, 249.
Praveen, R.S., Daniel, S., Prasada Rao, T., *Talanta*, 2005, 66, 513.
Renkecz, T., Mistlberger, G., Pawlak, M., Hováth, V., Bakker, E., *App. Mater. Interfaces*, 2013, 5, 8537.
Sadyrbaeva, T. Z., Russ. *J. Appl. Chem.*, 2013,86, 186.
Sahiner, N., Malci, S., Celikbiçak, O., Kantoglu, O., Salih, B., *Radiat. Phys. Chem.*, 2005, 74, 76.
Segatelli, M.G., Santos, V.S., Presotto, A.B.T., Yoshida, I.V.P., Tarley, C.R.T., *React. Funct. Polym.*, 2010, 70, 325.
Shao, D., Ren, X., Hu, J., Chen, Y., Wang, X., *Colloids Surf. A*, 2010, 360, 74.

Souza, M.O., Tarley, C.R.T., *Int.J. Environ. Anal. Chem.*,2009, 89,489.
Soylak, M., Kaya, B., Tuzen, M., *J. Hazard. Mater.*, 2007, 147, 832.
Soylak, M.,Tuzen, M., *J. Hazard. Mater.*, 2006,137, 1496.
Tajodini N., Moghimi, A., *Asian J. Chem.*, 2010, 22, 3335.
Tarley, C.R.T., Fernandes, F.F., Luccas, P.O., Segatelli, M.G., *Anal. Lett.*, 2011, 44, 216.
Tsukagoshi, K., Yu, K.Y., Maeda, M., Takagi, M., *Bull. Chem. Soc. Jpn.*, 1993, 66, 114.
Wu,G., Wang, Z., Wang, J., He, C., *Anal. Chim. Acta*, 2007, 582, 304.
Xu, L., Sun, J., Zhao, L., *Radiat. Phys. Chem.*, 2011, 80, 1268.
Yalçinkaya, Ö., Kalfa, O.M., Türker, A.R., *J. Hazard. Mater.*, 2011, 195, 332.
Ye, Y., Ali, A., Yin, X., *Talanta*, 2002, 57, 945.
Zhou, S., Xue, A., Zhao, Y., Wang, Q., Chen, Y., Li, M., Xing, W., *Desalination*, 2011, 270, 269.

In: New Developments in Analytical Chemistry … ISBN: 978-1-63463-427-4
Editor: Breanna Granger

Chapter 5

DEVELOPMENT OF A BIOSENSOR-BASED CARBON NANOTUBE PASTE (MWCT) -MODIFIED DNA ADSORBED ON INORGANIC MATERIAL FOR THE PHENOTHIAZINE DETERMINATION

***Guilherme B. Braga*[1,*], *Juliana de Fatima Giarola*[1], *Fernanda Midori de Oliveira*[3], *Emerson S. Ribeiro*[2,†], *César Ricardo Teixeira Tarley*[3,‡] and *Arnaldo C. Pereira*[1]**

[1]Departamento de Ciências Naturais, Universidade Federal de São João del Rei (UFSJ), Campus Dom Bosco, São João del Rei, MG, Brazil

[2]Departamento de Química Inorgânica, Instituto de Química, Universidade Federal do Rio de Janeiro (UFRJ), Rio de Janeiro, RJ, Brazil

[3]Departamento de Química, Universidade Estadual de Londrina (UEL), Centro de Ciências Exatas, Rodovia Celso Garcia Cid, Londrina, PR, Brazil

* Departamento de Ciências Naturais,Universidade Federal de São João del Rei (UFSJ),Campus Dom Bosco, Praça Dom Helvécio 74, Fábricas, 36301-160 São João del Rei, MG, Brazil.

† Departamento de Química Inorgânica, Instituto de Química, Universidade Federal do Rio de Janeiro (UFRJ), 21941-909 Rio de Janeiro, RJ, Brazil.

‡ Departamento de Química, Universidade Estadual de Londrina (UEL), Centro de Ciências Exatas, Rodovia Celso Garcia Cid PR 445 Km 380, 86050-482 Londrina, PR, Brazil. Corresponding author. Tel +55 43 3371 4366; fax +55 43 3371 4286; E-mail address: ctarleyquim@yahoo.com.br (C. R. T. Tarley).

ABSTRACT

An electrochemical method for phenothiazine (PTZ) determination, based on development of a sensor of carbon nanotube paste/DNA adsorbed on $SiO_2/Al_2O_3/Nb_2O_5$ (designated as SiAlNb), obtained by sol-gel process/mineral oil (60, 30, 10% w/w) was developed. The aim of this work was to evaluate the performance of the carbon nanotube sensor with DNA as catalyst for the voltammetric determination of phenothiazine. The modified electrode showed electrocatalytic activity towards the oxidation of PTZ. Experimental parameters were optimized and the best conditions were obtained: concentration of DNA adsorbed on the silica matrix was 0.5 mg mL^{-1}; in 0.10 M phosphate buffer solution (pH 6.5) using cyclic voltammetry. The sensitivity obtained with the proposed device for the phenothiazine determination, in the concentration range of 2.5 to 12.5 μmol L^{-1}, was about 1.5-fold in relation to carbon nanotube paste. Moreover, from the plot E_{pa} vs. pH, the E_{pa} for PTZ, shifted linearly (linear correlation of 0.9986) to more negative with increasing the solution pH, resulting in a slope of −58.41 mV/pH. This behavior shows, taking into account a Nernstian system, that the electrode reaction for PTZ takes place with a number of protons equal to the electrons. The determination of phenothiazine was carried out by differential pulse voltammetry (DPV) under the best conditions, pulse amplitude of 60 mV and scan rate of 5 mV s^{-1}. The analytical curve built by DPV was ranged from 0.083 up to 16.66 μmol L^{-1} ($r = 0.998$) with a sensibility 0.8313 μA μmol^{-1} L and precision < 2%. The limits of detection and quantification were found to be 0.023 μmol L^{-1} and 0.072 μmol L^{-1}, respectively. These data demonstrate that the proposed sensor is quite feasible for the determination of phenothiazine group.

Keywords: Inorganic matrix, carbon nanotube paste, DNA, phenothiazine

1. INTRODUCTION

The discovery of DNA electroactivity by 1960 (Palecek, 1960) has made the field of biosensors involving studies related to this biomolecule (Siddiquee et al., 2012) begin a frank expansion, with research to genetic analysis, elucidation of disease and even forensic and medical applications (Luo et al., 2013; Wang et al., 2011). In this context, various analytical techniques involving DNA determination, such as fluorescence (Xiang et al., 2013), electrochemical (Fojta et al., 2011), electrochemiluminescence (Su at al., 2011) chemiluminescence (Ye at al., 2011), surface plasmon resonance spectroscopy (Yu at al., 2009) and quartz crystal microbalance (Altintas et al., 2013) are employed. The use of electrochemical techniques emerges as an excellent opportunity for research with DNA (Du at al., 2009) mainly due to its simplicity, fast response, high sensitivity and selectivity, low cost, wide applicability and capability of miniaturization (Wang et al., 2011; Siddiquee et al., 2011).

DNA is a fundamental biomolecule, which stores genetic information, and thus is considered as important target molecules to interact with various characteristics. (Siddiquee et al., 2012). With this approach, the construction of biosensors based on DNA become relevant due to the electroactivity of DNA that resides at specific points of its structure; the nitrogenous bases (Meric at al., 2002). Thus for the determination of species in solution [15] that allow the interaction with DNA is necessary to consider the process of hybridization of this biological component, which by means of electrochemical techniques is evidenced by the amplification of the electrochemical signal of target species (Meric at al., 2002; Liu et al., 2010).

CNs (carbon nanotubes) possesses a special tube structure with unique properties such as good electrical conductivity, high chemical stability and extremely high mechanical strength (Ionescu et al., 2011; de Meneses et al., 2011). In addition, the subtle electronic behavior of CNs reveals that they have the ability to promote electron-transfer reaction and have a high electrocatalytic effect when used as electrode materials (Li et al., 2011; Beitollahi at al., 2011; Liu et al., 2011). These properties make CNs as a suitable candidate for the modification of electrodes as DNA. The use of DNA as the recognition element of affinity biosensors has been an increasing interest in research (McGown at al., 1995; Millan et al., 1992; Palecek, 1996). DNA recognition layers can be employed for detecting analytes interacting with such layers. Small molecules as drugs can be detected via changes in their response (associated with their interaction with the recognition layer), through intrinsic changes in the DNA signal (due to such association) or via the displacement of certain markers (Palecek, 1996; Pandey et al., 1994).

Bearing in mind the aforementioned comments, this study shows the development of a biosensor for determining PTZ, which consisted of a composite based on the multiwall carbon nanotubes (MWCN) paste and an inorganic material based on silica - $SiO_2/Al_2O_3/Nb_2O_5$ (SiAlNb). The electrodes modified with inorganic materials practically form a new class of chemically modified electrodes. However, the inorganic materials can be incorporated in the composites, generally with the aim of increasing the stability of the electroactive species immobilized. Several inorganic matrices have been used (Walcarius, 1998; Walcarius, 1999; Pereira et al., 2004) as modifying agent conventional electrodes. These include silica gel obtained by sol-gel process, metal oxides and phosphates, γ-alumina, zeolites, and other types.

A variety of materials based on silica together the chemistry of their surface implies a great potential for the application of these species in electrochemistry. Some of the properties of this material such as adsorption capacity, acid-base chemical, oxide metallic dispersed on the silica matrix, thermal stability, can advantageously be exploited, for example, the accumulation of electroactive species prior to electrochemical detection (Walcarius, 2001a; Walcarius, 2001b; Budziak et al., 2003). Furthermore, the silica can be grafted with a variety of functional groups leading to a considerable enrichment of their surface properties. The high surface area synthetic silica, when combined with their surface chemistry, makes this material useful as supports for various catalysts.

Recently it was demonstrated the use of material SiAlNb for the preconcentration of Zn (II) and Cd (II) in solution (Costa et al., 2011; Tarley et al., 2010). Also, the use of the SiAlNb in development of electrochemical biosensors for amitriptyline and promethazine determinations has been described in the literature. (Marco et al., 2013a; Marco et al., 2013b). The use of SiAlNb material becomes interesting, because it displays great potential to prepare chemically modified electrodes.

Phenothiazine compounds among its main features are considered tranquilizers, antihistamines, antipsychotics, sedatives, anesthetic, analgesic, which have as a starting material for synthesis; phenothiazine (PTZ) (Mahajan et al., 2012). In the 50s the discovery of tranquilizers properties of phenothiazine derivatives compounds by means of synthesis of chlorpromazine was one milestone, which allowed the use of this drug to elucidate their actions in the activity in the central nervous system (Ali et al., 2013).

This study aimed to develop an analytical method using an electrochemical DNA biosensor based on immobilized matrix modified silica ($SiO_2/Al_2O_3/Nb_2O_5$), aiming at the determination of phenothiazine solution.

Therefore the aim of this study was to develop and characterize a DNA modified multiwall carbon nanotubes paste electrode and to determine (PTZ) in solution, which served as a “sensor base” for the determination of phenothiazine compounds in raw material and/or coated tablets.

2. EXPERIMENTAL

2.1. Apparatus

The electrochemical measurements were performed with an Autolab potentiostat/galvanostat (PGSTAT 12, Eco Chemie, the Netherlands) interfaced with a personal computer for data acquisition and potential control. The experimental conditions were controlled with General Purpose Electrochemical System (GPES) software. A conventional three electrode cell was used at 25 ± 1°C. An Ag/AgCl/KCl (3.0 mol L^{-1})/KCl (3.0 mol L^{-1}) electrode, a platinum wire, and the MWCN-SiAlNb-DNA were used as the reference, auxiliary and working electrodes, respectively. The prepared electrodes were characterized by FT-IR model 8300 from Shimadzu (Tokyo, Japan). The KBr pellet technique was used to characterization of materials. The scanning electron micrograph images were taken using a JEOL model JSM 300 (Tokyo, Japan). The Al_2O_3 and Nb_2O_5 contents in the material were determined by using energy dispersive X-ray fluorescence analysis (EDFRX) on a model EDX 800 HS from Shimadzu (Tokyo, Japan). The analysis of the specific surface area (S_{BET}) of the materials were performed on equipment Quantachrome Model Nova 1200e (Boynton Beach, Florida, USA) and determined by the BET (Brunauer, Emmett and Teller) multipoint method by submitting the samples to previous activation at 250ºC in vacuum for 4 h. The measures of average pore size and average pore volume were obtained through the Barrett-Joyner-Halenda (BJH) method.

2.2. Reagents

Multiwall carbon nanotubes (MWCN) were acquired from CN Corporation Ltd (Incheon, Korea) with diameter of 10-40 nm and length of 5-20 μm. Herring sperm deoxyribonucleic acid (DNA), phenothiazine, *piperazine-1,4-bis [2-ethanesulfonic acid*] (PIPES), 4- [2-hydroxyethyl] piperazine-1-ethanesulfonic acid (HEPES), tris [hydroxymethyl] aminomethane (Trizma), sodium hydroxide, nitric acid and mineral oil were acquired from Sigma-Aldrich® (St. Louis, MO, USA). Potassium chloride were obtained from Sial® (São Paulo, SP, Brazil). Dibasic sodium phosphate was acquired from Vetec® (Rio de Janeiro, RJ, Brazil). The material $SiO_2/Al_2O_3/Nb_2O_5$ (designated as SiAlNb), obtained by sol-gel process, was prepared according to previously described procedure (Costa et al., 2011; Tarley et al., 2010).

All other chemicals were of analytical grade with the highest purity available. Water was distilled and purified using a Millipore Milli-Q Plus system (Bedford, MA, USA).

2.3. Preparation of the Biosensor Using $SiO_2/Al_2O3/Nb_2O_5$ (SiAlNb) and Multi-wall Carbon Nanotubes

For the immobilization of the biological component (DNA) on the surface of the matrix sol-gel silica (SiAlNb), 300 mg of this material was added in 60 mL of a solution containing 0.5 mg mL^{-1} of DNA prepared in 0.10 mol L^{-1} phosphate buffer pH 6.5. This mixture was stirred for 15 min and allowed to stand by 12 h for adsorption of biological component in the matrix. The mixture was filtrated and then dried at room temperature. This mixture was called SiAlNb-DNA.

After the drying process, it was prepared a MWCN paste containing material SiAlNb-DNA prepared previously. This material (60 mg) consisted of a mixture 1:1 (w/w) between the SiAlNb-DNA and CNs in mineral oil. This final mixture (MWCN-SiAlNb-DNA) was introduced into the lower cavity of the glass tube, which was employed as working electrode.

2.4. Preparation of Reference Solutions

Stock standard solutions of PTZ were prepared by dissolution of the drug in water to obtain a final concentration of 0.01 mol L^{-1}. This standard solution was diluted to give concentrations between 2.5 and 20.0 μmol L^{-1} of active pharmaceutical ingredient. All of these solutions were stored at -20ºC in the absence of light. Working standard solutions were prepared daily by diluting the stock solutions to an appropriate concentration with 0.1 mol L^{-1} phosphate buffer pH 6.5.

2.5. Recovery Test

The recovery test of PTZ in spiked human urine in two levels (1.0 and 15.0 μmol L^{-1}) was examined by using standard addition method. An aliquot of a human urine sample was diluted two-fold with 0.1 mol L^{-1} phosphate buffer solution at pH 6.5 and the mixture was spiked with a solution containing an accurately measured amount of each level of PTZ. The urine sample was filtered using a syringe filter of 0.45-μm pore size before the experiment at the DNA-modified electrode.

3. RESULTS AND DISCUSSION

3.1. Characterization of Studied Sensor and Biosensor

The results of the analysis of EDFRX revealed that were incorporated about 19.9 wt% of Al_2O_3 and 20.9 wt% of Nb_2O_5 in the SiAlNb. The analysis obtained by the BET method showed the specific surface area of 306 m^2 g^{-1} to SiAlNb. The BJH method displayed average pore size of 30.6 Å showing that the materials is mesoporous and the values found for the average pore volume was of 0.026 cm^{-3} g^{-1}. These values are close to those reported in literature (Zheng at al., 2009). These values reflect also a good accessibility to active sites, fundamental characteristic of adsorption processes of biological material, as noted in item 3.2.1.

(a)

Figure 1. (Continued).

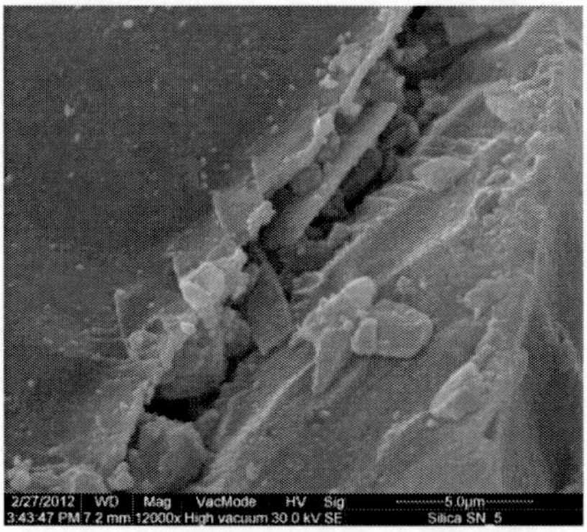

(b)

(c)

(d)

Figure 1. Scanning electron microscopy images of the MWCN (a), $SiO_2/Al_2O_3/$ Nb_2O_5 (SiAlNb) (b), DNA (c) and MWCN-SiAlNb-DNA (d).

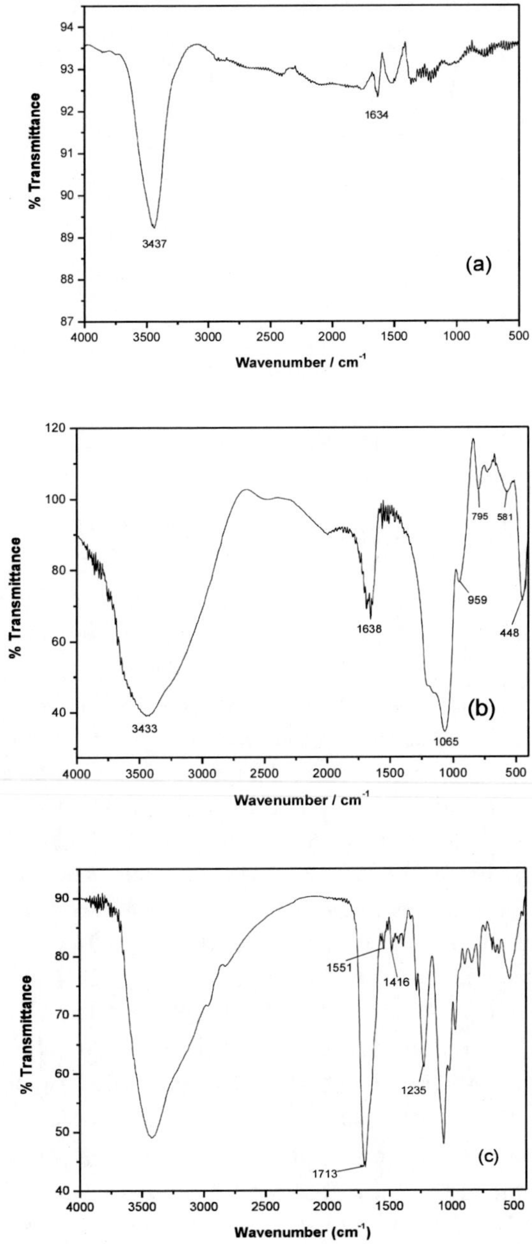

Figure 2. (Continued).

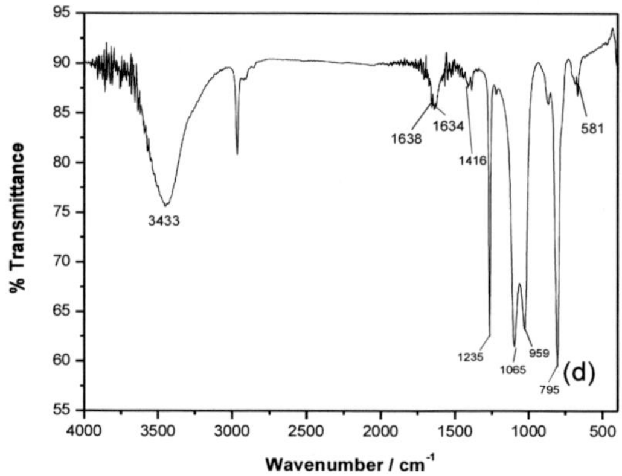

Figure 2. FT-IV spectra of the MWCN (a), $SiO_2/Al_2O_3/Nb_2O_5$ (SiAlNb) (b), DNA (c) and MWCN-SiAlNb-DNA (d) (Marco et al., 2013).

Figure 1 shows the SEM images of the MWCN (a), $SiO_2/Al_2O_3/Nb_2O_5$ (SiAlNb) (b), DNA (c) and MWCN-SiAlNb-DNA (d). MWCN present a low degree of entanglement due to hydrophobic forces between nanotubes, while SiAlNb, as expected, presents a flat surface very common in mixed oxides grafted onto silica surface; on the other hand, the surface of DNA is very rough. The surface of composite MWCN-SiAlNb-DNA is very irregular and present morphological aspect as resulted of synergic interaction among the materials, thus showing modification in the biosensor. As observed, the carbon nanotubes are impregnated in the matrix of SiAlNb and DNA.

Figure 2 shows the infrared spectra of the MWCN (a), SiAlNb (b), DNA (c) and MWCN-SiAlNb-DNA (d). For MWCN, the band at 3437 cm^{-1} is attributed to stretching vibrations of OH, indicating the existence of carboxyl groups on the external surface of the MWCN or water adsorbed. The signal at 1634 cm^{-1} can be assigned to stretching vibration of C = C, as well as adsorbed water molecules (Buhani et al., 2010). For SiAlNb material a typical broad band at 3433 cm^{-1} was observed, which is attributed to stretching vibration of O-H groups involved in hydrogen bonds ($\nu_{O\text{-}H}$). In similar way to MWCN, the band at 1638 cm^{-1} can be attributed to adsorbed water molecules, but also to νSi-O-Si absorption (Aronne et al., 2007). The absorptions between 1065 and 1163 cm^{-1} are attributed to νSi-O absorption from the silica support (Silva et al., 2009). The band at 959 cm^{-1} probably is assigned to stretching and bending modes of the Si-OH groups ($\nu_{Si\text{-}OH}$), while in 795 and 959 cm^{-1} are relative to stretching mode of Si-O-Si. The bands at around 581 and 448 cm^{-1} are attributed to the deformation of Si-O-Si (δSi-O-Si). For the DNA, the FT-IR spectra shows a band at 1713 cm^{-1} related to the stretching of carboxyl group (C = O) from guanines bases of nucleic acid. The absorption bands at 1652 and 1625 cm^{-1} can be assigned to stretching vibration of C = C, C = O and C = N (Wang et al., 2001), while the stretching vibration of C-N and C-C from ring of puryne base was observed at 1551 cm^{-1} and 1416 cm^{-1} is related to C-H and deformation of N-H from imidazole ring. The band at 1235 cm^{-1} is assigned to vibration of pirimidine ring in nucleic acid (Neault et al., 1999). According to FT-IR results, it is clearly confirmed the successful of chemical modification of SiAlNb by DNA, since the characteristics bands in Figure 2d were observed in the spectra of MWCN-SiAlNb-DNA.

3.2. Optimization of the Biosensor Preparation

The influence of the biological material and different matrices were studied to analyze PTZ employing the cyclic voltammetry (CV). Table 1 presents the results obtained for the electrodes of graphite (GRT), graphite modified with SiAlNb (GRT-SiAlNb), graphite modified with SiAlNb and DNA (GRT-SiAlNb-DNA), multi wall carbon nanotubes (MWCN), multi wall carbon nanotubes modified with SiAlNb (MWCN-SiAlNb) and multi wall carbon nanotubes modified with SiAlNb and DNA (MWCN-SiAlNb-DNA).

The results presented in Table 1 showed the importance of simultaneous use of MWCN modified with SiAlNb and the biological component DNA (MWCN-SiAlNb-DNA). This association promoted the catalytic oxidation of PHZ increasing the sensitivity. The increased sensitivity afforded by MWCN-SiAlNb-DNA can be attributed to the low charge transfer resistance of the MWCN and a better dispersion and fixation of DNA in silica of Nb_2O_5 promoted by acid groups present, which immobilize better basic groups of molecules of DNA. In addition, there may be a preconcentration by adsorption of the PTZ in the surface of the DNA through a strong interaction; such higher affinity is attributed to the intercalative binding of the drug with the surface-confined DNA layer (Liu et al., 2011; Wang et al., 1996).

3.3. Influence of Concentration of DNA in SiAlNb Matrix

The influence of DNA concentration in the SiAlNb matrix is a factor very importance since the amount of compound on the electrode surface determines the magnitude of intensity of the anodic peak current (I_{pa}) for the analyte of interest, i.e. the sensitivity. So, different concentrations of DNA (0.25 to 1.5 mg mL^{-1}) were used in the immobilizing step of the DNA on the surface of the matrix. In this study, we evaluated the *sensitivity* by CV for PTZ in the range of concentration (2.5 to 12.5 µmol L^{-1}) in each prepared biosensor. Figure 3 shows the results obtained for each concentration of DNA. The concentration of 0.5 mg mL^{-1} presented the best result with *sensitivity* of 0.27 µA μmol^{-1} L.

Table 1. Sensibilities of proposed electrodes for PTZ analysis by cyclic voltammetry (The analytes and 0.5 mg mL^{-1} DNA were prepared in 0.1 mol L^{-1} phosphate buffer pH 7.0)

Sensor composition	Sensitivity (µA/µmol L^{-1})
GRT	0.189
GRT-SiAlNb	0.157
GRT-SiAlNb-DNA	0.222
MWCN	0.0784
MWCN-SiAlNb	0.165
MWCN-SiAlNb-DNA	0.267

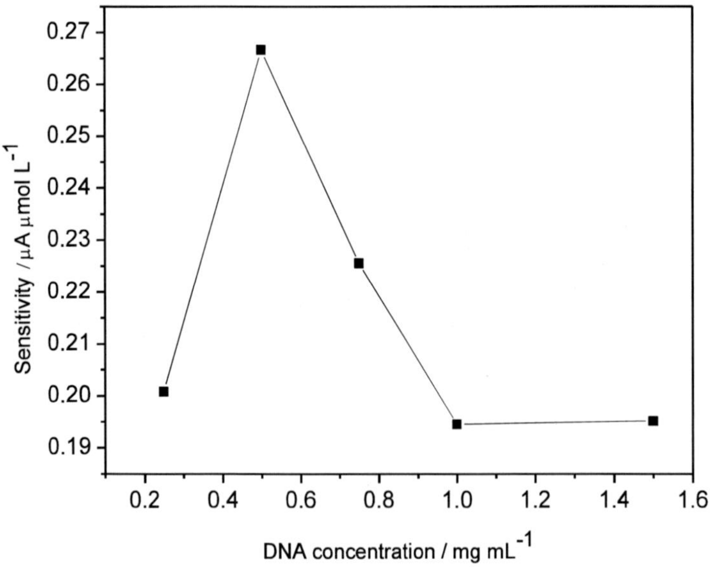

Figure 3. Sensitivities obtained in different concentration of DNA immobilized in the SiAlNb matrix.

3.4. Electrochemical Behavior of MWCN-SiAlNb-DNA Biosensor

To evaluate the electrocatalytic activity of the biosensor MWCN-SiAlNb-DNA were obtained cyclic voltammograms (CVs) in the range of 0.3 to 1.3 V vs Ag/AgCl/KCl (3.0 mol L^{-1}). Figure 4a shows CVs obtained with sensors prepared with MWCN and MWCN-SiAlNb. In both cases, no faradaic process in range of studied potential was observed. However, the biosensor MWCN-SiAlNb-DNA presented two peaks detected in 0.80 and 1.10 V (Figure 4a). These peaks in 0.80 and 1.10 V can be attributed to oxidation of the DNA, residual bands of guanine and adenine, respectively (Zheng et al., 2009; Sun et al., 2008, Brett et al., 1998; Mello et al., 2009). Figure 4b illustrates the CVs obtained in the range of 0.0 to 0.5 V vs Ag/AgCl/KCl (3.0 mol L^{-1}) with the same sensors/biosensors studied in Figure 4a, but in the presence of analyte (7.5 μmol L^{-1} PTZ). The presence of DNA promoted a significant increase of signal (anodic peak current). Figure 4c shows the CVs obtained by MWCN-SiAlNb-DNA in the absence and presence of analyte. It was only observed an electrochemical signal employing MWCN-SiAlNb-DNA in the presence of PTZ.

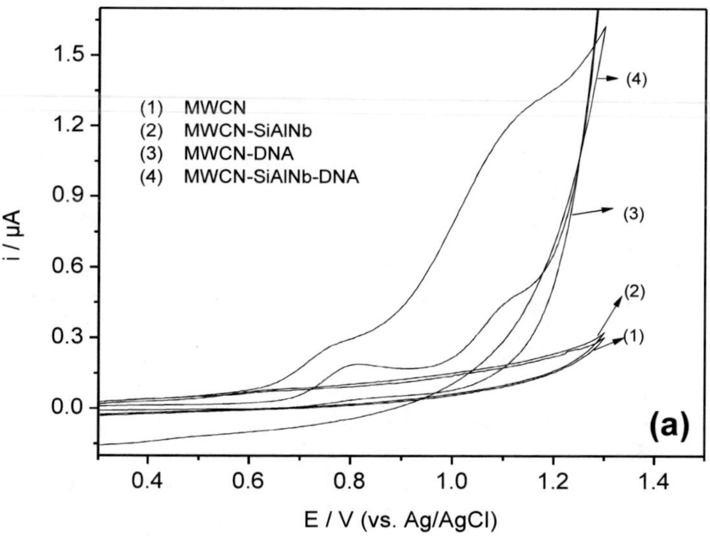

Figure 4. (Continued).

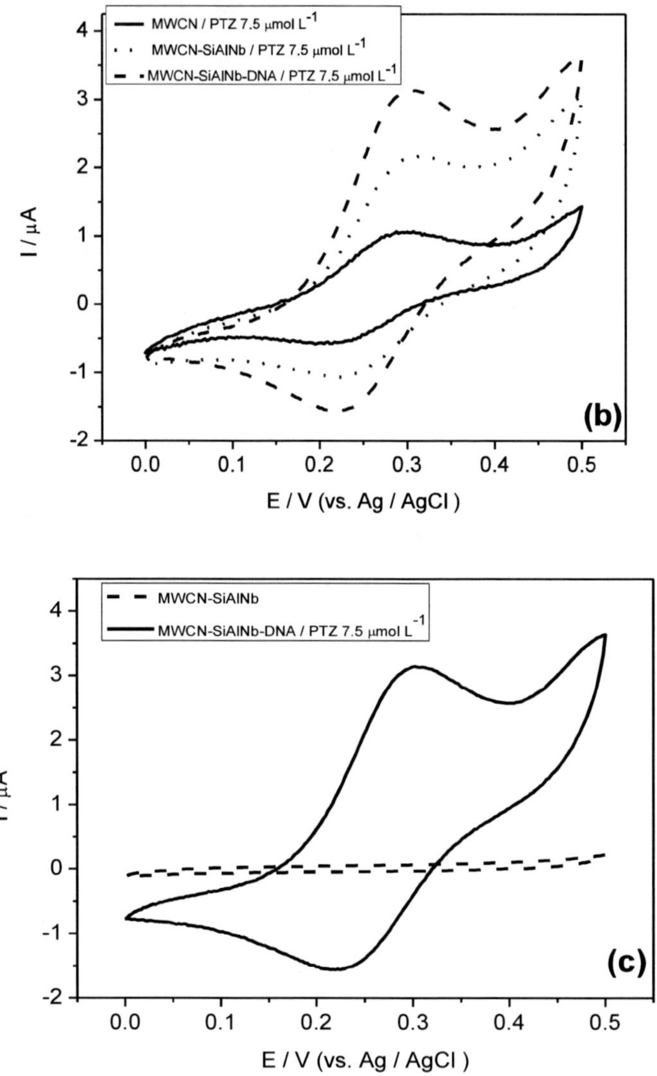

Figure 4. (a) Cyclic voltammograms obtained with the following biosensors and sensors (1) MWCN, (2) MWCN-SiAlNb (3) MWCN-DNA and (4) MWCN-SiAlNb-DNA in the absence of PTZ; (b) Electrocatalytic effect of sensors and biosensors using 7.50 μmol L^{-1} PTZ and (c) Biosensor of MWCN-SiAlNb-DNA in the absence and presence of 7.50 μmol L^{-1} PTZ. Experiments were performed in 0.1 mol L^{-1} phosphate buffer pH 7.0 at $\nu = 10$ mVs^{-1} in the potential range from 0.3 to 1.3 V vs. Ag/AgCl/KCl (3.0 mol L^{-1}) in case (a) and the potential range from 0.0 to 0.5 V vs. Ag/AgCl/KCl (3.0 mol L^{-1}) in cases (b) and (c).

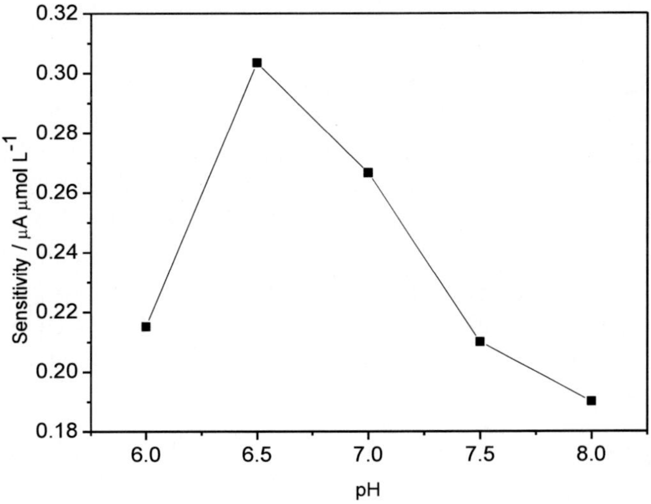

Figure 5a. pH influence in response of biosensor in 0.1 mol L^{-1} phosphate buffer at 10 mVs^{-1}.

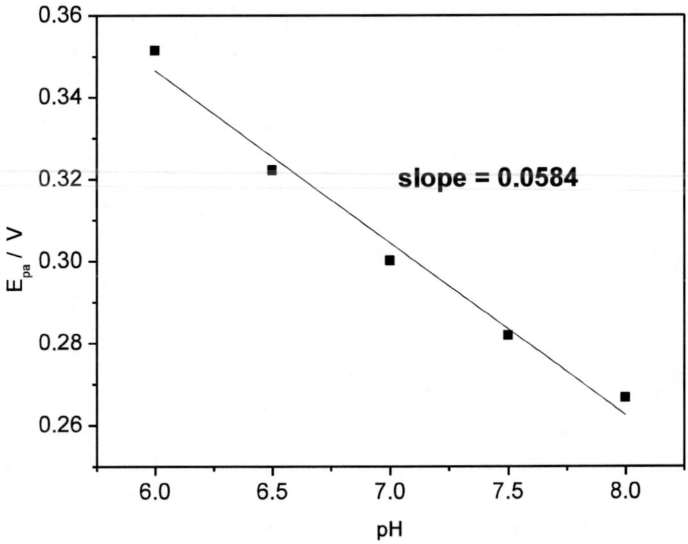

Figure 5b. Relationship between the anodic peak potential and pH obtained in a solution at 7.50 µmol L^{-1} of PTZ employing the proposed biosensor.

3.5. pH Influence

The effect of influence of pH on the response of biosensor was performed in phosphate buffer 0.1 mol L^{-1} with pH values ranged 5.0 to 7.5. The results showed a maximum of I_{pa} when the buffer solution pH was 6.5 (Figure 5a).

Another evidence observed from this study concerns the behavior of the anodic peak potential as a function of pH. From the plot E_{pa} vs. pH, the E_{pa} for PTZ, shifted linearly (linear correlation of 0.998) to more negative with increasing the solution pH, resulting in a slope of -58.41 mV/pH (Figure 5b). This behavior shows, taking into account a Nernstian system, that the electrode reaction for PTZ takes place with a number of protons equal to the electrons.

3.6. Buffer Solution Influence

The following buffer solutions were used to available the best response for the oxidation of PTZ employing the biosensor MWCN-SiAlNb-DNA: HEPES, phosphate, Tris-HCl and Pipes. All were prepared in 0.1 mol L^{-1} concentration at pH 6.5. Figure 6 presents the values of sensitivities and the best results were obtained with phosphate buffer. This may be related to smaller size of phosphate ions that allows higher mobility due higher electron transfer between the MWCN-SiAlNb-DNA biosensor and the buffer solution. This phenomenon keeps the electroneutrality of the electrode, which resulted in a higher oxidation current for PTZ.

3.7. Buffer Solution Concentration

The following buffer concentration (pH 6.5) were used to available the best response for the oxidation of PTZ employing the biosensor MWCN-SiAlNb-DNA: 0.03, 0.05, 0.10, 0.20, 0.30 mol L^{-1}. The highest values of sensitivities were obtained in concentration of 0.10 mol L^{-1} (Figure 7). With the increase of the ionic strength until 0.10 mol L^{-1}, the sensitivity was gradually increased, but decreased by further increase of the salt concentration. This effect probably is attributed to the denaturation of the biological material. The concentration of phosphate buffer equal to 0.10 mol L^{-1} was used in the next experiments.

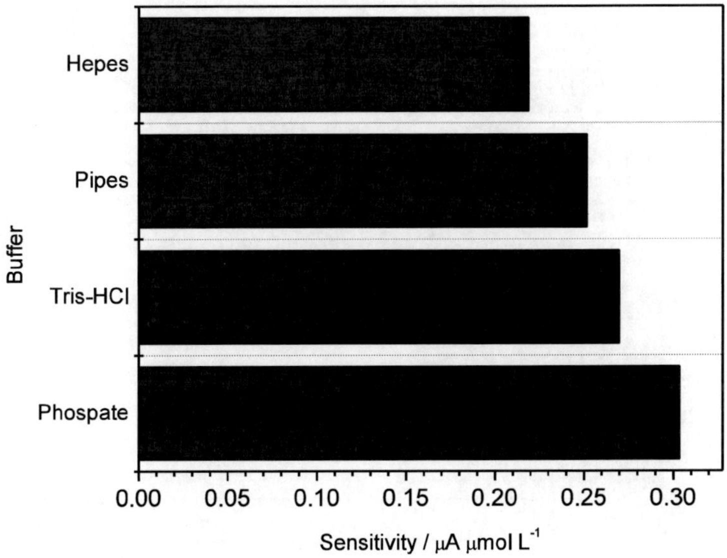

Figure 6. Buffer influence in response of biosensor at 0.1 mol L^{-1} pH 6.5 at 10 m Vs^{-1}.

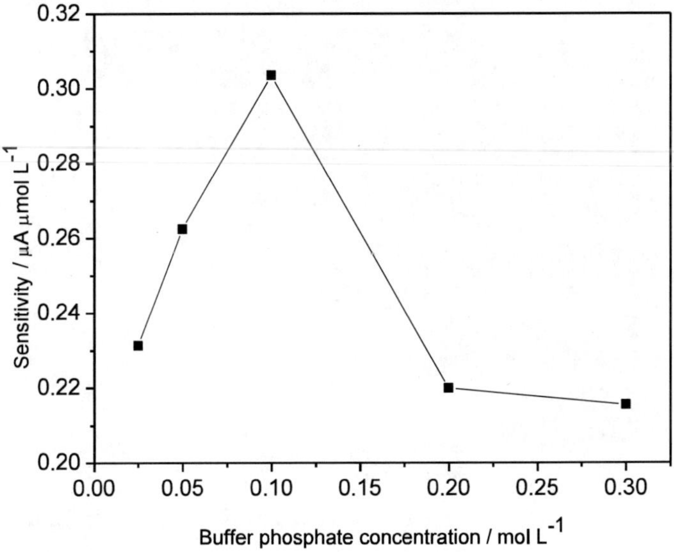

Figure 7. Influence of concentration of phosphate buffer in response of biosensor in pH 6.5 at 10 mVs^{-1}.

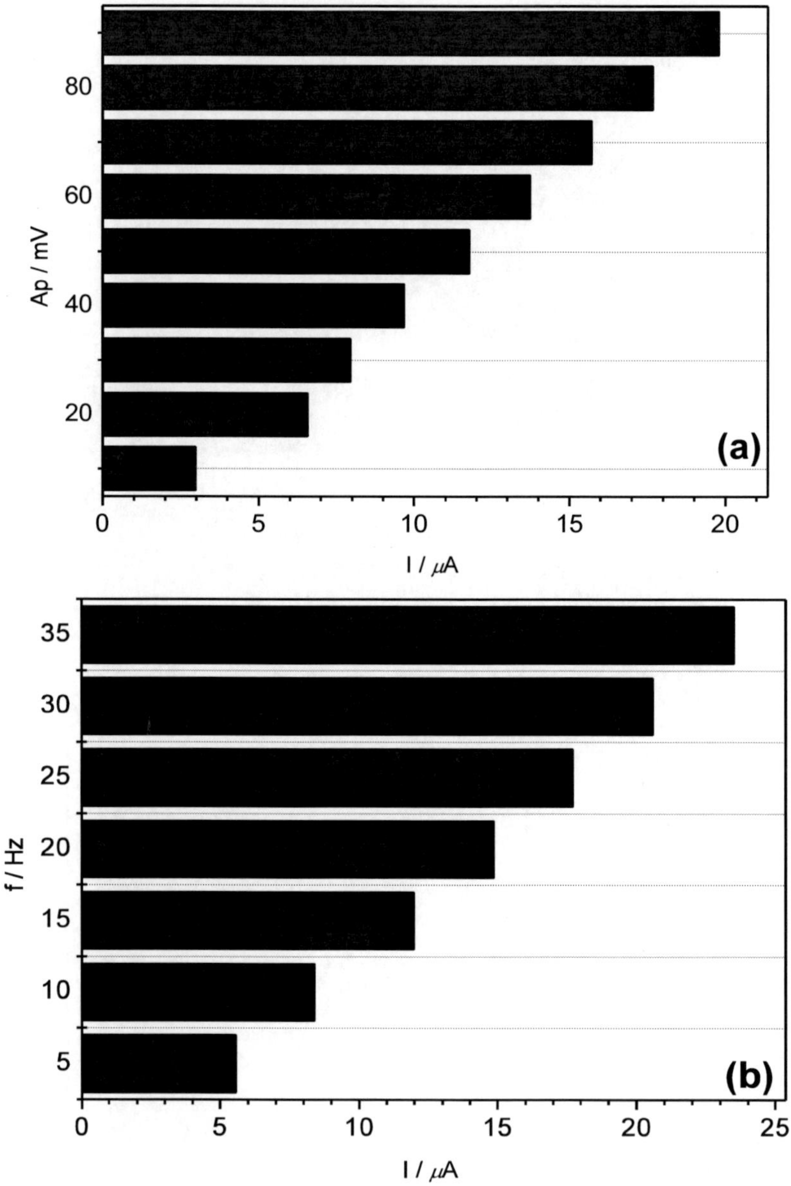

Figure 8. Influence of (a) pulse amplitude; Ap; and (b) frequency; f; on the response of the proposed biosensor employing VOQ. Measurements performed in the presence of 7.5 mmol L^{-1} PTZ prepared in phosphate buffer solution 0.1 mol L^{-1}, pH = 6.5.

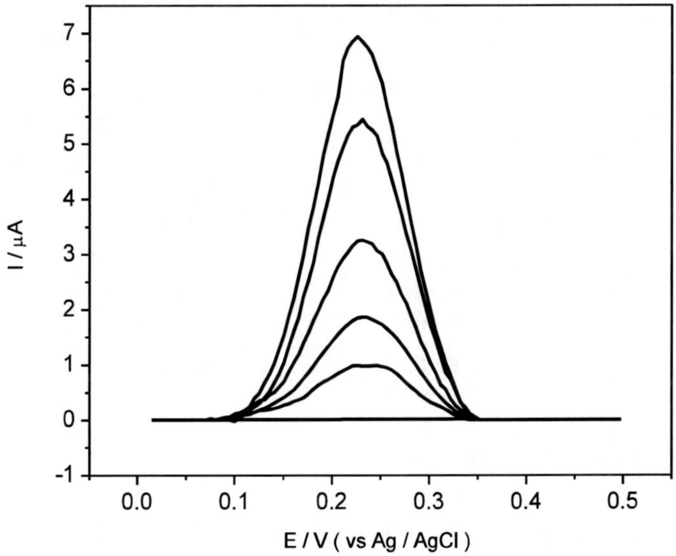

Figure 9. Square Wave Voltammograms obtained to PTZ. Measurements performed employing the MWCN-SiAlNb-DNA under experimental conditions and operating optimized in solution 0.1 mol L^{-1} phosphate buffer pH 6.5.

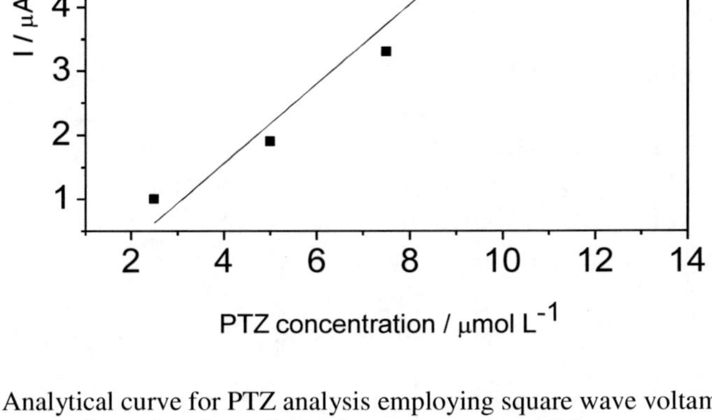

Figure 10. Analytical curve for PTZ analysis employing square wave voltammetry. Measurements performed under optimized experimental conditions.

3.8. Choice of Electroanalytical Technique

The choice of the electrochemical technique to be used for the determination of an analyte is of fundamental importance since analytical parameters such as LOD, LOQ, sensitivity, selectivity, precision, accuracy, among others are dependent employed technique. It was evaluated two electroanalytical techniques: square wave voltammetry and differential pulse voltammetry for the determination of PTZ using the working electrode MWCN-SiAlNb-DNA. Analytical curves were constructed using each technique optimized to evaluate which of these techniques would be used to determine PTZ in real samples. The technique of differential pulse voltammetry presented better sensitivity and it was used to determine PTZ (Table 2).

3.9. Square Wave Voltammetry (SWV)

The SWV presented as important parameters: frequency (*f*) and pulse amplitude (*Ap*); and thus, these parameters were optimized in order to obtain higher value of anodic peak current (I_{pa}) and an ideal voltammetric profile, i.e., exhibiting no distortion or enlargement peak

Thus, Figure 8 illustrates the effects of pulse amplitude and frequency on the response of the proposed biosensor.

After the optimization of operating parameters; f = 15 Hz and Ap = 30 mV; an analytical curve was obtained for determination of PTZ through the SWV technique, which is shown in Figure 9.

3.10. Differential Pulse Voltammetry (DPV)

The DPV consists in the same principle of the VOQ; i.e., generation of pulses along the applied potential aiming at an increasing the sensitivity of the proposed method. To optimize the electrochemical method is recurring consider the following parameters involved in this technique: pulse amplitude (*Ap*) and scan rate (*v*). In this context as to ensure a higher value oxidation peak current (I_{pa}) and an ideal voltammetric profile, the parameters were evaluated as shown in Figure 11.

After the optimization of operating parameters; $v = 5$ mvs^{-1} e $Ap = 60$ mV; an analytical curve was obtained for determination of PTZ through the DPV technique, which is shown in Figure 12.

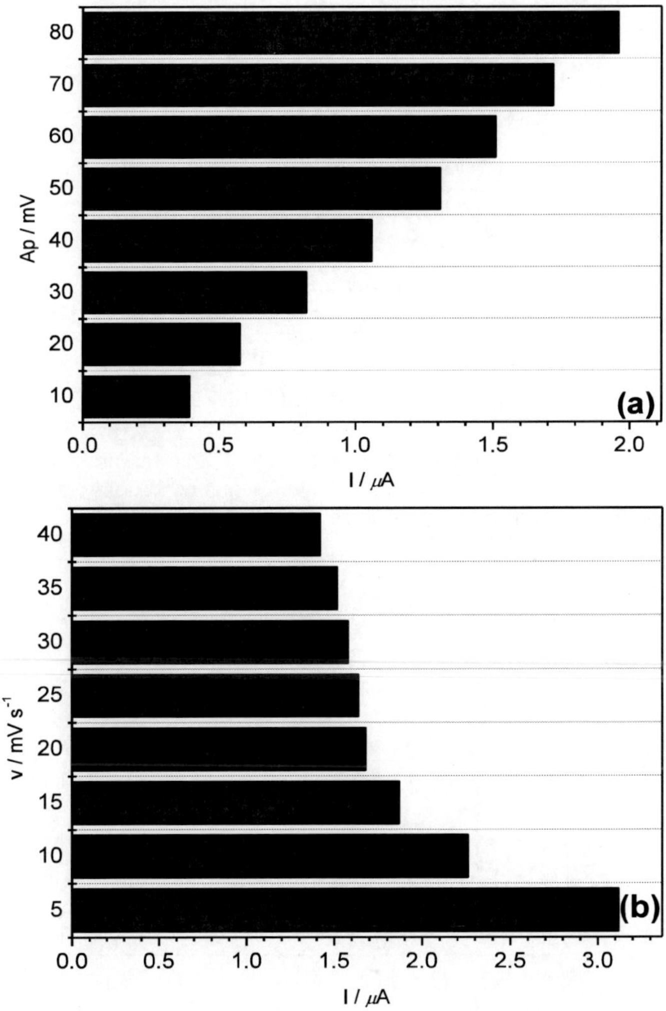

Figure 11. Influence of (a) pulse amplitude; Ap; and (b) scan rate ν; on the response of the proposed biosensor employing DPV. Measurements performed in the presence of 7.5 mmol L^{-1} FTZ prepared in phosphate buffer 0.1 mol L^{-1}, pH = 6.5.

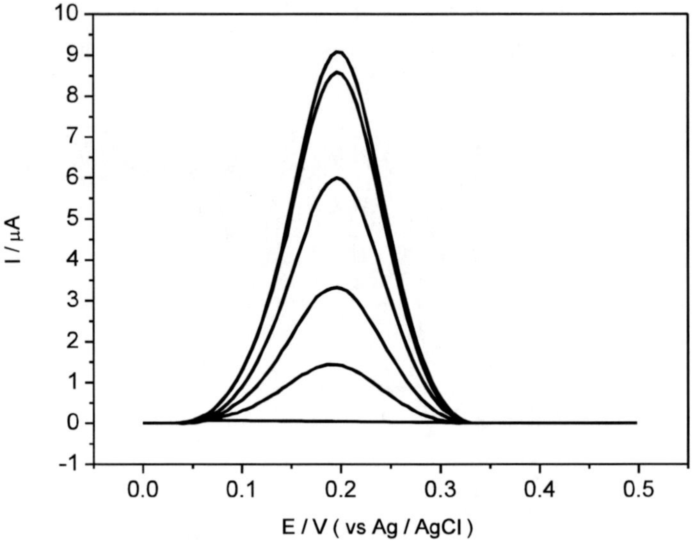

Figure 12. Differential Pulse Voltammetry obtained to PTZ. Measurements performed employing the MWCN-SiAlNb-DNA under experimental conditions and operating optimized in solution 0.1 mol L−1 phosphate buffer pH 6.5.

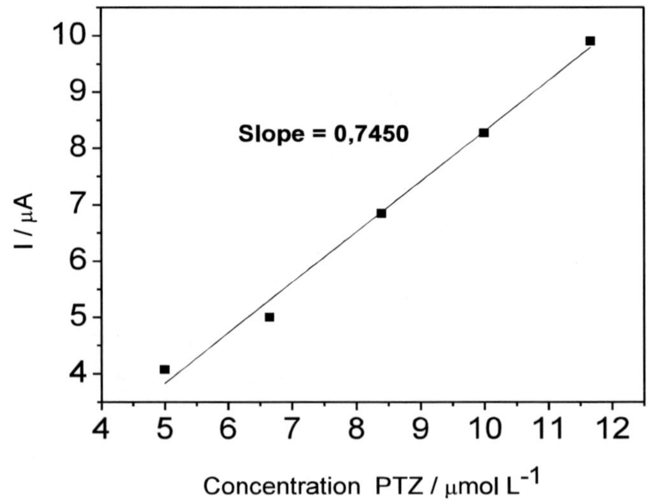

Figure 13. Analytical curve for PTZ analysis employing differential pulse voltammetry. Measurements performed under optimized experimental conditions.

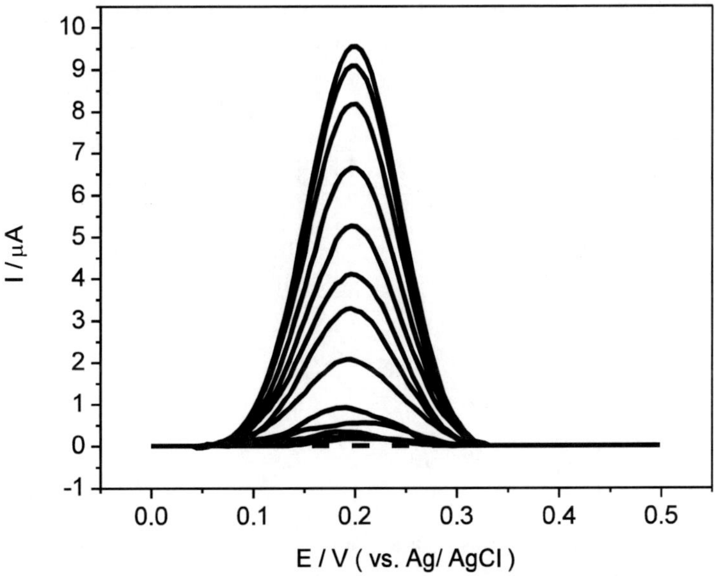

Figure 14. Differential Pulse Voltammetry obtained to PTZ employing the proposed biosensor in a concentration range of 0.083 mmol L^{-1} to 16.66 mmol L^{-1}. Measurements performed employing the MWCN-SiAlNbp-DNA under experimental conditions and operating optimized in solution 0.1 mol L^{-1} phosphate buffer pH 6.5.

3.11. Preparation of a Final Analytical Curve

After optimization of the experimental and operational parameters, which are presented in Table 2, a final calibration curve using the proposed method for determination of FTZ solution biosensor was constructed.

Figure 14 shows the differential pulse voltammograms obtained at different concentrations of PTZ, which showed a linear response in the range studied, with the correlation coefficient (R = 0.99). Figure 15 shows the calibration curve obtained.

As from the analytical curve were extracted some figures of merit (Table 3), such as sensitivity, detection limit (LOD), quantification limit (LOQ) and linear response range of the analytical method; wherein for the calculation of LOD and LOQ the following equations were used according to IUPAC LD = 3 s/b; LQ = 10 s/b.

Table 2. Results of the optimization of the experimental and operational parameters

Parameters Evaluated	Optimized Conditions
Sensor Configuration	MWCN-SiAlNb-DNA
DNA concentration	0.5 mg mL^{-1}
pH	6.5
Electrolyte Composition	Phosphate buffer solution
Electrolyte Concentration	0.1 mol L^{-1}
Electrochemical Technique	DPV

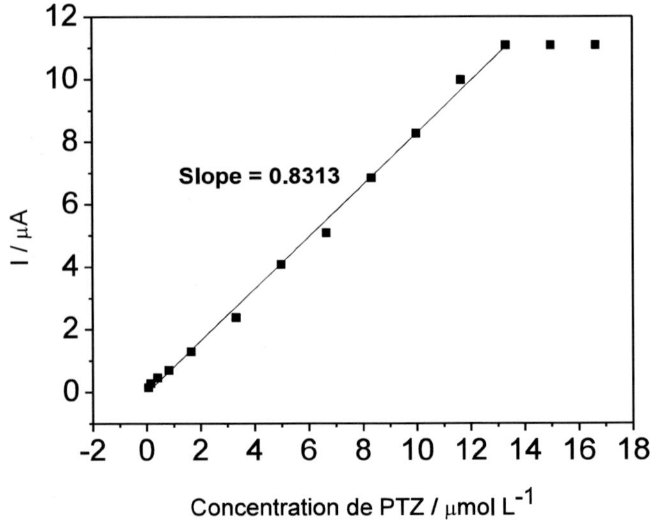

Figure 15. Analytical curve for PTZ analysis employing differential pulse voltammetry in a concentration range of 0.083 mmol L^{-1} to 16.66 mmol L^{-1}. Measurements performed under optimized experimental conditions.

3.12. Final Analytical Conditions

After optimization of the electroanalytical system proposed, it was verified that the best working conditions were: composition of the biosensor: MWCN-SiAlNb-DNA, DNA solution employed in this step biocatalyst immobilization onto the surface of inorganic material equal to 0.5 mg mL^{-1}, pH 6.5 equal to the working set in phosphate buffer in a concentration optimized and equal to 0.10 mol L^{-1}. After evaluation of the operational parameters of the different electrochemical techniques studied, we found that the proposed biosensor exhibited the best analytical response when measurements were performed by differential pulse voltammetry.

3.13. Interference Study

A variety of substances contained in urine, such as ascorbic acid and amino acids, can react with DNA and thus interfere the detection of phenothiazine. Therefore, signal response of 5.0 μM PTZ was measured at the present DNA-modified electrode in the presence of several possible interfering species at concentrations close to their physiological levels and the results are listed in Table 4. The presence of interfering species in the detection of PTZ gives rise to the anodic current response increase less than 10% at the DNA-modified electrode while more than 13% at the bare carbon nanotube paste electrode.

This decreased interference effect observed in the present DNA-modified electrode is likely due to the following reason.

The negatively charged phosphate backbones of DNA prevent the diffusion of negatively-charged interfering species into the DNA-modified electrode via electrostatic repulsion. Finally, the interfering species cannot be intercalated into the DNA duplex and thus cannot be preconcentrated into the DNA-modified electrode. The present interference study implies that the present method based on the DNA-modified electrode allows highly selective determination of PTZ without an additional membrane on an electrode surface for blocking the permeation of interfering species (Hwang et al., 2011).

3.14. Real Sample Analysis

To demonstrate the practical usage of the present method, the recovery test has been carried out for human urine sample spiked with two levels of PTZ concentration. The fresh urine sample was two-fold diluted with 0.10 M phosphate buffer at pH 6.5 and then analyzed by the standard addition method. As observed from Table 5, good results were obtained (> 93%), thus confirming the interference-free applicability of method. This indicates that the present method based on the DNA-modified electrode can be applied to the PTZ analysis in real sample such as human urine.

Table 3. Analytical parameters obtained from the final analytical curve

Figures of merit	Results
Linear response range	0.083 to 16.330 μmol L^{-1}
Linear equation	I = 0.8313 [PTZ] + 0.2250
Sensitivity	0.8313 μA/μmol L^{-1}
LOD	0.023 μmol L^{-1}
LOQ	0.072 μmol L^{-1}

Table 4. Interference study at the bare electrode (MWCN paste electrode) and the MWCN-SiAlNb-DNA modified electrode

Substance	Concentration (mM)	MWCN paste[a]	MWCN-SiAlNb-DNA modified electrode[a]
Alanine	1.00	1.18	1.08
Serine	1.00	1.13	1.04
Ascorbic acid	0.50	1.29	1.10
Glutamic acid	0.50	1.17	1.06
Methionine	0.10	1.25	1.05

[a] Relative Response (anodic current response of 5.0 μM PTZ + interfering substance / anodic current response of 5.0 μM PTZ without interfering substance).

Table 5. Determination of PTZ in urine sample

Sample	Phenothiazine added (μmol L^{-1})	Phenothiazine found (μmol L^{-1})[a]	Recovery (%)
1	1.00	1.07 ± 0.04	107
	15.0	13.9 ± 1.5	93

[a] Results are expressed as mean value ± standard deviation based on three replicates (n = 3).

CONCLUSION

A new, simple and inexpensive analytical method was developed for the potential determination of PTZ in raw material and pharmaceutical formulation, for example. Furthermore, this work has demonstrated the feasibility of developing sensors based on DNA for the determination of phenothiazines class, which can it be widely used as an antihistaminic for the symptomatic relief of hypersensitivity reactions and for enhancing the analgesic, anesthetic and sedative effect of other drugs.

The developed method proved to be selective, linear, precise and accurate. In addition, this method may be extended to other applications with some modifications proving to be a valuable alternative to analyses that employ conventional methods.

ACKNOWLEDGMENTS

The authors are grateful to Fundação de Amparo à Pesquisa do Estado de Minas Gerais (FAPEMIG), Conselho Nacional de Desenvolvimento Científico e Tecnológico (CNPq), Coordenação de Aperfeiçoamento de Pessoal de Nível Superior (CAPES) FAPERJ (Fundação Carlos Chagas Filho de Amparo à Pesquisa do Estado do Rio de Janeiro), Fundação Araucária do Paraná and INCT/Bioanalítica for financial support and for granting research fellowships, and the student of scientific initiation Felipo Doval Rojas Soares of IQ/UFRJ for the preparation of the material SiAlNb. This work is a collaboration research project of members of the Rede Mineira de Química (RQ-MG) supported by FAPEMIG.

REFERENCES

Ali, M., S., Al-Lohedan, H. A., *J. Mol. Liq.* 2013. 117. 283.

Altintas, Z., Tothill, I. E., *Sens. Act., B*, 2013. 169. 188.

Aronne, A., Marenna. E., Califano. V., Fanelli. E., Pernice. P., Trifuoggi. M., Vergara., A., *J. Sol-gel Sci. Technol.* 2007. 43. 193.

Beitollahi, H., Raoof, J. B., Hosseinzadeh, R., *Electroanalysis*, 2011. 23. 1934.

Brett, A. M. O., Macedo., T. R. A., Raimundo., D., Marques., M. H., Serrano., S. H. P., *Biosens. Bioelectron.* 1998. 13. 861.

Budziak, D., Silva., E. L. S., Campos., S. D., Carasek, E., *Microchim. Acta* 2003, 141. 169.

Buhani, Narsito, Nuryono, Kunarti, E. S., *Desalination* 2010. 251. 83.

Costa, L. M., Ribeiro., E. S., Segatelli., M. G., Nascimento., D. R., Oliveira., F. M., Tarley., C. R. T., *Spectrochim. Acta, Part B* 2011. 66. 329.

Du, P., Li. H., Cao, W., *Biosens Bioelectron.* 2009. 24. 3223.

Fojta, M., Kostecka., P., Pivonková., H., Horáková., P., Havran, L., *Current Anal. Chem.* 2011. 7. 35.

Hwang, R. Y., Xu., G. R., Han, J., Lee., J. Y., Choi, H. N., Lee, W. Y., *J Electroanal. Chem.* 2011. 656. 258.

http://www.sciencedirect.com/science/article/pii/S1572665711004103 - bb0200

Ionescu, M. I., Zhang, Y., Li., R., Sun, X., Abou-Rachid, H., Lussier, L. S., *Appl. Surf. Sci.* 2011. 257. 6843.

Li, H. B., Pan, L. K., Lu, T., Zhan,Y. K., Nie, C. Y., Sun, Z., *J. Electroanal. Chem.* 2011. 653. 40.

Liu, S., Liu. J., Han, X., Cui, Y., Wang, W., *Biosens. Bioelectron.* 2010. 25. 1640.

Liu, X. G., Peng, Y. H., Qu, X. J., Ai, S. Y., Han, R. X., Zhu, X. B., *J. Electroanal. Chem.* 2011. 654. 72.

Luo, C., Tang, H., Cheng, W., Yan, L., Zhang, D., Ju, H., Ding, S., *Biosens. Bioelectron.* 2013. 48. 132.

Marco, J. P., Borges, K. B., Tarley, C. R. T., Ribeiro, E. S., Pereira, A. C., *Electroanal. Chem.* 2013a, 704. 159.

Marco, J. P., Borges, K. B., Tarley, C. R. T., Ribeiro, E. S., Pereira, A. C., *Sens Act B,* 2013b. 177. 251.

Mahajan, S., Mahajan, R. K., *J. Colloid. Int. Sci.,* 2012, 387. 194.

McGown, L. B., Joseph, M. J., Pitner, J. B., Vonk, G. P., Linn, C. P., *Anal. Chem,* 1995. 67. 663A-668A.

Mello, L. D., Ribeiro, E. S., Kubota, L. T., Elmroth, S. K. C., Pereira, R. M. S., *Biometals* 2009 22. 385.

De Menezes, V. M., Rocha, A. R., Zanella, I., Mota, R., Fazzio, A., Fagan, S. B. *Chem. Phys. Lett.* 2011. 506. 233.

Meric, B., Kerman, K., Ozkant, D., Kara, P., Erensoy, S., Akarca, U.S., Mascini, M., Ozsoz, M., *Talanta*, 2002. 56. 837.

Millan, K. M., Spurmanis, A. J., Mikkelsen, S. R., *Electroanalysis* 1992. 4. 929.

Neault, J. F., Riahi, H. A. T., *Biophys. J.* 1999. 76. 2177.

Palecek, E., *Nature*, 1960. 188. 656.

Palecek, E., *Electroanalysis* 1996. 8. 7.

Pandey, P. C., Weetall, H. H., *Anal. Chem.* 1994. 66. 1236.

Pereira, A. C., Kubota, L. T., *Quim. Nova* 2004. 27. 725.

Siddiquee, S., Yusof, N. A., Salle, A. B., Tan, S. G., Bakar, F. A., *Microchim Acta*, 2011. 172. 357.

Siddiquee, S., Yusof, N. A., Salle, A. B., Tan, S. G., Bakar, F. A., *J. Solid State Electrochem.* 2012. 16. 273.

Silva, A. L. R., Sousa, K. S., Germano, A. F. S., Oliveira, V. V., Espinola, J. G. P., Fonseca, M. G., Airoldi, C., Arakaki, T., Arakaki, L. N. H. *Colloids Surf., A* 2009. 332. 144.

Su, B., Tang, D., Tang, J., Li, Q., Chen, G., *Anal. Biochem.* 2011. 417. 89.

Sun, W., Li, Y., Yang, M., Liu, S., Jiao, K., *Electrochem. Commun.* 2008. 10. 298.

Tarley, C. R. T., Ávila, T. C., Segatelli, M. G., Lima, G. F., Peregrino, G. S., Scheeren, C. W., Dias, S. L. P., Ribeiro, E. S., *J. Braz. Chem. Soc.* 2010. 21. 1106.

Walcarius, A., *Electroanalysis* 1998. 10. 1217.

Walcarius, A., *Anal. Chim. Acta* 1999. 384. 1.

Walcarius, A., *Electroanalysis* 2001. 13. 701.

Walcarius, A., *Chem. Mater.* 2001. 13. 3351.

Wang, J., Rivas., G., Cai., X., Shiraishi, H., Farias, P. A. M., Dontha, N., Luo, D., *Anal. Chim. Acta* 1996. 332. 139.

Wang, L., Chen, X., Wang, X., Han, X., Liu, S., Zhao, C., *Biosens. Bioelectron* 2011. 30. 151.

Wang, Z., Liu, D., Dong, S., *Bioelectrochemistry* 2001. 53. 175.

Xiang, D. S., Zhai, K., Wang, L. Z., *Analyst*, 2013. 138. 5318.

Ye, S., Li, H., Cao, W., *Biosens. Bioelectron.* 2011. 26. 2215.

Yu, L. P., Sun, Y. Z., Zhao, Z. X., *Curr. Pharmaceut. Anal.* 2009. 5. 112.

Zheng, Y., Yang, C., Pu, W., Zhang, J., *Microchim. Acta* 2009. 166. 21.

INDEX

C

D

E

F

G

K

L

M

N

O

P

Q

R

S

T

U

V

W

Y

Z